Core Samples

CORE SAMPLES

A Climate Scientist's Experiments in Politics and Motherhood

Anna Farro Henderson

UNIVERSITY OF MINNESOTA PRESS

MINNEAPOLIS • LONDON

Anna Henderson was a fiscal year 2019 recipient of an Artist Initiative grant from the Minnesota State Arts Board. The creation of this book was supported by the voters of Minnesota through a grant from the Minnesota State Arts Board, thanks to a legislative appropriation from the arts and cultural heritage fund.

Published by the University of Minnesota Press
111 Third Avenue South, Suite 290
Minneapolis, MN 55401-2520
http://www.upress.umn.edu

ISBN 978-1-5179-1604-6 (pb)

A Cataloging-in-Publication record for this book is available from the Library of Congress.

Printed in the United States of America on acid-free paper

The University of Minnesota is an equal-opportunity educator and employer.

32 31 30 29 28 27 26 25 24   10 9 8 7 6 5 4 3 2 1

Field Notes

Prologue

Science as Story

The first tool I acquired as a scientist was a Rite in the Rain notebook. It was small and yellow and had a waxy feel. Paddling a canoe or swimming to collect vials of lake water, I felt the work pass from my body into my pencil and onto the page. I could write on a boat with waves sloshing over me. While the pages got soaked, my notes and sketches remained the sharp metallic gray of graphite. Muddy fingerprints the color of chocolate milk layered another story on top of water depths and GPS coordinates. My tool kit expanded to include a rock hammer, an ice ax, a U.S. Senate badge, a breast pump, and twenty-four-hour access to the Minnesota State Capitol. But it was always my notebook that made me invincible.

My notes informed hypotheses about the natural world. But they went beyond this, too. When I trained to do research in a submarine or moved an expedition's campsite because of wildfire, my hair went unwashed, my clothing collected tears, love affairs sparked, and grudges took hold. Living science, embodying the process of inquiry, engaged all my senses. In my field notes, I documented the landscape around mountain lakes as well as the shock of jumping in with a small bottle and swimming with icebergs to collect a sample. While the data from this water offered a puzzle piece of science, the experiences were also part of a larger story I was trying to make sense of: Myself.

As I made poems out of days of fieldwork and laboratory analysis, I came to see art as a slant form of the scientific process. Here too, I made observations, posed questions, and tested ideas. While some people see art and science as opposites, for me, they form a braided river, each strand and flow an approach to wonder. For my PhD and postdoctoral research in geology, I tried to constrain myself to research alone. But it felt incomplete. The traditional vehicles for publishing science research—peer-reviewed journals—were inadequate for exploring or sharing the full spectrum of insights that came from my scientific work. Through artistic expression, I could delve into uncharted spaces.

My research focused on climate change. I "read" lake mud, collected over thousands of years, as archives of ancient periods of warming. The work could inform our predictions of the future. It was the mid- to late aughts. But no matter how many data points I stacked up, I didn't have an avenue for engaging with the political dialogue. When efforts to set federal climate policies failed, I didn't get it. From my lab bench, I could not understand why the science was not enough to drive political change. I wanted to make a difference—to do something.

When I became a mother, time caught up with me. As I confronted my body and the world in new ways, the need to write became fundamental. The thought of explaining myself to the baby raised in me an awareness of being in the depths of climate research but having no connection to direct action. No one could hear me shouting in a laboratory.

I hung up my lab coat to work in Washington, D.C., as an environmental policy adviser to Senator Al Franken of Minnesota. I aimed to be a two-way translator—as if all we needed to solve climate change were better explanations. Most policies had a foundation in science or technology. While much of the policy I worked on was unrelated to my research, my background allowed me to navigate and parse details. I could talk through the specific requirements of federal fracking regulations or make recommendations related to genetically modified food labeling concerns. My fluency in technical details gave me power—even when I was the only woman in the room.

In Congress, research findings were reduced to talking points: a statistic on the size of the dead zone in the Gulf of Mexico, a map of mineral resources, and glossy information sheets listing sterile factoids. I missed science as open-ended discovery. In policy debates, science was just one of many variables considered, and, generally, it was not central to negotiations. Stripped of nuance and broader context, our relationship to science was no longer dynamic. Even when raised, the topic of climate change seemed to fizzle and fall out of the conversation.

I found that amid the congressional gridlock, a battle raged to shape the societal narrative on which action hinged. Proposed bills, hearings, negotiations, threats, and speeches all fed competing stories. Advocates struggled to integrate climate into the main arena of political concern. Framing policies as protections—of oceans, of the atmosphere, of biodiversity—made these issues seem separate from us. Our deep interconnectedness with the forests, rivers, and landscapes that shape our lives wasn't coming through.

Engaging with nature could seem like a choice—a luxury even. Tangible and immediate anxieties threatened communities. Job losses, the cost of health care, and the need for affordable energy were more urgent. While economic and public health outcomes were tied to air and water quality, pollution control efforts posed their own challenges. Making the required changes could destabilize businesses and communities. I saw how the proposed solutions raised questions beyond those science could answer. Most actions required trade-offs and risks, and while science might inform legislation, value judgments drove votes.

It took the implementation of a new environmental regulation for me to understand fully why data alone were not enough to convince people to take action on climate change. I'd gone to the Minnesota State Capitol to serve as an adviser to Governor Mark Dayton. In the land of ten thousand lakes, entire regions were unsafe for fishing or swimming because of contamination from agriculture and other industries. Yet the pushback against regulation was huge, even within our own administration.

Traveling the state to host town hall meetings on water policy, I listened to people from the prairies, conifer forests, and farmlands. They talked about their connection to the land and how proposed policies might affect their lives. I heard several times that it was just part of living in farm country not to be able to drink the water. I understood then that we don't live by science. We live by our stories.

Our climate crisis is not just a technical challenge. It is a human and cultural problem. To build public will, and participation, we need language that makes sense of the rapid rate and profound scale of environmental change. We need to see ourselves in relationship with our environment—and with each other. We need stories to integrate the impacts of environmental changes on our lives and to share what these mean with each other. Stories are how we shape and change culture.

In this book I share my story as a scientist in research, political advising, art making, and motherhood. The essays that follow take different forms, from meditations to user manuals. One is a packing list, another a lactation diary. This book is not, primarily, an explanation of science concepts or a guide to political procedure. It doesn't make political arguments. Nor is it an exposé or a comprehensive documentation, though known political figures appear. My stories bring readers into the daily rhythms and intimacies of research expeditions and political negotiations.

The stories I tell in this book illuminate flaws in research, the halls of government, and the role of science in policymaking. These cracks offer an invitation for engagement. Revealing the humanity of our institutions allows us to make changes in the story.

Irreversible decisions are being made about the management of natural resources, and our future hangs in the balance. It has never been so critical for the public to engage with the scientific and political processes that govern our lives. The decisions we make now will affect our health, well-being, and ways of life for generations. By sharing my lived experience as a scientist and policy adviser, I aim to open the doors of science and politics. Walk with me through these messy, and very human, institutions.

1. First Experiments

To initiate a relationship to the landscape, I learn the names of plants as though they are secret passwords. The ripple effects of our warming climate are boundless. I arm myself with the theory of change that by reconstructing ancient periods of climate warming, we can better understand the potential impact and scale of current and future changes. To read the past, I collect lake mud from drill rigs or whatever floating platforms I can cobble together. I get to know trees by their needles and the texture of their bark. To study ecosystem changes over millennia, I learn to identify pollen under a microscope and chemical markers of plants with high-tech instruments. I piece together ancient periods of climate warming that affected lakes and rivers, forest fires and droughts. The ongoing tension of being a geologist is never having all the pieces to complete the picture. I work toward discoveries that can lead only to more questions.

What If?

◆

Coming of Age as a Scientist

By simply accepting the invitation to contemplate
the "what if" we unleash our superpower on the very
nature of possibility itself, oftentimes resulting
in the seemingly miraculous.

LEVAR BURTON

My first big experiment, one with consequence, is drinking
untreated water. I am six years old and living in Dharamshala,
India. To make water safe, I can boil it or filter it or dissolve iodine
in it. I carry a metal canteen of treated water to kindergarten, but
it never lasts. I get thirsty. I sit cotton-mouthed in class. Mouth
dry and lips chapped, I will have a long walk up the mountain
after school. Considering that my peers drink untreated water, I
guess that I can too. I know tiny creatures live in my gut. I figure
I've lived in India long enough to have the kind of creatures that
can handle the water.

One day I raise my hand in class, and, like any other kid, I skip
to the courtyard fountain. The water tastes cool and delicious. I
picture the amoebas wriggling down my throat. The experiment
works. I am fine.

My family left Providence, Rhode Island, in 1985. Starting in
Hong Kong, we spent close to a year traveling by bus, train, and
car. My parents, Canadian and American academics, had studied
and lived with Tibetan communities before. My older brother
was born in Dharamshala in the late seventies. While my parents
say they are raising us Jewish, at bedtime we read stories about

the Monkey God and from kid versions of the Mahābhārata and the Rāmāyana. My body has come to know the rhythm of circumnavigating temples to the click of prayer beads and chants of *om mani padme hum.*

Each morning in India, I hike from the grounds of the Dip Tse Chok Ling Monastery where we live through the forest to the bus stop. Monkeys watch from high branches. I'm not pleased about having to keep my Sacred Heart School uniform tidy. My mom grows exasperated. "Could you not get scratches on your face?" she asks. But the trees offer sticks to bend and sharpen into bows and arrows. My pleated kilt, tucked-in shirt, and clip-on tie constrain free movement. The dark green of my skirt is something from a swamp. I realize, though, that my uniform is all that says I belong. My siblings and I are the only white children at school. I can clasp hands and run with classmates who speak Hindi, all of us united by our red-and-gold pins with the motto *Cor unum.* Through the forest, we reach the bus stop, and our parents launch us the rest of the way down the mountain by pushing the bus until the pistons fire.

Living abroad demands I pay attention. Invites wonder. Cumin on sizzling lamb fat cooked by a Uyghur street vendor. The tang of yogurt made from the milk of a *dhzo,* a mix of yak and cow, shared by nomads. The crunch of shrimp chips given to me by Tibetan teens I meet in an open-air market who want to hang out. The cardamom sweetness of halwa, a carrot pudding, delivered by the blind waiter in the crowded teahouse. These moments seem miraculous. But occasionally, we can't find enough food to buy. Not all of what I see, or experience, is so sweet or so satisfying. The trip teeters between the excitement of adventure and the buzzing of an electric fence. Some moments invite concern, terror even.

When we travel on the Tibetan Plateau, we come across a field of boulders. I ask where the rocks come from. Earthquakes, my parents say. The ground below us transforms into a sleeping beast. The beast doesn't wake that day, but it roars several months later in Dharamshala. A community higher up the mountain, accessed by a stone staircase through cedar forest, crumbles. We once had dinner with a family who lived there and kept rabbits. They had let us play with their animals, perhaps they also fed them to us.

By Qinghai Lake, I watch goats on a hillside ripe with wild-flowers. I giggle at—or with—the herd. A boy comes and whistles for the animals. The first to reach him has his neck slit. That is dinner.

My fear mingles with curiosity: rocks move. Usually, they move on a timescale we can't fathom. But the land has its own agency. I see myself in dialogue as I dig, run, and play in forests, plains, and courtyards.

I pump water to wash clothing by hand. Our bathroom at the monastery is a shared concrete outhouse. In Lhasa, we hovered over a hole cut from the floor on the second story of a hotel. Bark-ing dogs punctuate the smell of ammonia and shit.

I understand that in my hometown the pipes that run through the walls of our house connect to networks of drainage below the streets, allowing us to live with an ease that distances us from the need to gather food or water or deal with our waste.

Somewhere along the way, my parents throw their wedding rings off a mountain. They tell us they are no longer together. But we are too far from home for this to make sense.

When we return to Providence, the reasons to study my surround-ings change, but the necessity is no less pressing. Once home, my parents actually split up. This means we four kids are divided and then merged with new households as each parent remarries. I gain three more siblings plus a kind of uncle in a stepsibling's other parent. That makes for a total of seven kids, five adults, and three homes. It also makes for short fuses. I learn, like a sailor watching the currents and clouds, to anticipate, avoid, and mit-igate storms.

I tell romantic stories about my large family to friends in my children's theater group: feet thumping up and down the stairs, doors swinging open, the landline ringing off the hook, skate-boards slapping pavement, and the dinner table crowded with friends. The hypnosis of perpetual motion is one way to deal with the chaos. My other armor is to preemptively announce that I am weird and my family fucked up, conflating the two.

To escape, I dive into books—an entry to imagination. When Roald Dahl's tonsils are removed without anesthesia in *Boy*, I throw the book across my room, screaming. Reading is a full-bodied act, and the escape lasts beyond the time spent in the pages. Running outside in a rainstorm, I am in *Haroun and the Sea of Stories*. Home alone, eyes closed, I walk through the house as Helen Keller. I inhabit characters like a ghost possessing them.

Despite my connection to characters in books, onstage, with the theater company, I'm not particularly good at acting. My focus is outward. I am too aware of the cast and audience to be present with myself, never mind myself-as-someone-else. I want to be a star, a beautiful princess, of course. I know, every girl wants that. And, because we all take turns, I eventually play the lead in *Twelfth Night*. I love reading Shakespeare aloud with the cast. I memorize my lines. I worship the director who exclaims, "Jiminy Cricket!" She leads us in vocal warm-ups and meditations on outer space. When my mom forgets or is late to pick me up, the director takes me to her apartment as though it's no inconvenience. The theater is my third space, a place away from home and school. Here work and community center on playfulness.

But I don't enjoy being in the spotlight—what I actually love is operating the stage lights. Climbing a ladder to reach the tiny hobbit hole, I can look down on the audience and performers. The expanse of knobs and switches fascinates me. I build a map in my mind of all the ways lighting can create a scene. During performances, I run to the pay phone and order pesto pizza with sun-dried tomatoes to be delivered to the bottom of my ladder. With lights, I frame drama and set the mood. I build an atmosphere of delight or doom, chaos or serenity. While I am gregarious, I feel most myself behind the scenes.

"I'm going to be a writer when I grow up," I tell my dad. I have just written an elaborate death scene sitting on my closet floor. When we visit his family in British Columbia, he takes us on long swims across bays so cold I feel the space between every finger and toe. Writing is like that. I find flow. For brief moments the world makes sense.

"Writers starve. You have to get a real job when you grow up," my dad responds. Then, as if remembering something, he says, "Artists starve too."

By my parents' measure I am not one of the smart ones. Learning to read doesn't come easy. Maybe a late bloomer—I had lost some of my hearing to tubes that stopped almost-constant earaches. Teachers express concerns about a learning disability, but this never gets sorted out. When I get into the public gifted program where my older siblings went for middle school, my parents instead keep me in the neighborhood.

"You are my last hope for a scientist in the family," my Grandma May says as her blue Buick crosses over the sparkling Fraser River in Vancouver. She doesn't mean a research scientist—she means a Christian Scientist. Grandma May had tuberculosis as a teenager, and for what were supposed to have been her last years, her parents sent her to London to train as a concert pianist. Music was her dream. She delivers her stories of England with utter seriousness, but they play like fairy tales. It rained frogs, ghosts ran onto the balcony at teatime, sun rays fell upon her in a garden after she converted to Christian Science, and she was healed.

In her pocket diary from that time, a leather-bound book that fits in the palm of my hand, she wrote on the first day of 1927, "Lesson with Mr. Matthey soon after returning. Marvelous. On Mozart Fantasia. Remembering most his remark about every moment's value. The hand of time passes on—in music—in life." *The hand of time passes on in music and in life.* This stays with me.

"Red for *fleishig,* blue for *milchig,* and white for *pareve,*" my Grandma Fanny instructs me on keeping distinct sets of dishes, silverware, and tablecloths for a kosher household. This is the only Yiddish she passes on from her mother tongue—the language she spoke before kindergarten.

I feel closest to Grandma Fanny, so I ask what she thinks happens when we die. We are walking the mile to shul for service. I want to imagine the *what if?* What if when she passes, I can still feel her presence?

"When you die, you are dead," she says.

"But what about God?" I hold her gaze, her beautiful blue eyes insectile beneath her glasses.

"God?" She laughs her laugh of ten thousand bells. "There's no god in Judaism." She had gone to her dad's hometown in Romania when she was thirteen and met her extended family. The older generations had fled pogroms in the Ukraine. By the time my mother was born, those cousins, aunts, and uncles Grandma had met had died in concentration camps or been otherwise lost to the war. She doesn't talk much about the Holocaust. Mostly, she focuses on the future. "Make sure you have six children, one for every million Jews killed," she tells me.

I don't dream about being a mother or getting married. I try to picture myself in college or owning a house, but the vision hovers fuzzy and gray.

For my grandparents, religion makes sense of the world. Orthodox Judaism and Christian Science, with or without a god, provide them with frameworks for sharing life—in their households and in their communities. But when these frameworks smash together, the spirit becomes malicious. Both families opposed my parents' marriage. My uncle even sat shiva, as if my mother had died.

For a while, after the divorce, my mom takes us to synagogue on Saturdays, and my dad takes us to church on Sundays. While I admire the domed architecture and music, an invisible force separates me from the congregations. In synagogue, I find excuses to slip out of the service. In Sunday school, I stage an intractable battle with the teacher, insisting that bugs and trees and animals have souls. But my parents' goal is not to build connection—it is a show of power, a competition.

Other than their personal, and ongoing, conflict, my parents are neither religious nor political. They focus their intensity in service of their careers. I suspect they garner more belonging from work than from family.

The work for my high school classes does not hook me or bestow a sense of purpose. Generally, school is not about learning. I go to school for the gathering—it is a social venue.

To understand how others keep from feeling totally out of control, I study my peers. A friend juggles at kids' birthday parties and steals street signs at night. A friend ties a raincoat around her waist as an amulet. A friend wears a shirt continuously until the constellation of holes makes a tectonic gash. A friend wants to come to my house but never lets me inside hers. A friend is convinced she is Mediterranean royalty. Friends find logic or solace in Narnia, summer camps, Rastafarianism, witchcraft, eating disorders, their dad's drug recovery, or their own experience getting high.

While I observe, forming my own hypotheses, science in my Providence public high school classroom is as good as dead. As in, the fume hoods are defunct—no experimenting. Teachers leave us to teach ourselves. One year I don't have a biology teacher, and when a substitute finally comes, she has no experience with students. I use the class period to write notes—the medium of friendships, courtships, and dissolutions.

Once I throw off the curve of a chemistry exam by scoring double the average. This plants a seed. I can sit in a library for hours learning on my own, and I don't *not* enjoy it. A classmate exclaims, "How is that possible? Anna is always cavorting." It is true: I cavort in the hallways making out with boyfriends. I cavort after school with friends. I cavort on weekends rambling around the city. I join the field hockey team that never wins a game, and I cavort making up silly or dirty cheers. I set up a tent in the yard to cavort by living outside for a summer, stretching an extension cord from the house to play cassette tapes. Nothing I do projects a scholarly vibe.

Toward the end of high school, a beloved older sibling becomes scattered and abruptly returns from college. They stay in the psychiatric hospital where H. P. Lovecraft visited his parents. I too feel the world as uncanny.

Crises smack like dominos. Just as senior year starts, a custody battle over a younger sibling pulls me into court. I spend a day in a hot, crowded hallway dreading what the judge might ask. Luck on my side, time runs out. But, in the fallout, my mom and her second family move as far away as they can possibly go.

My father tells me that some people can handle more than others, and if they can, they should. I am the least of anyone's concern. My final field hockey season is lost to mononucleosis. I muddle through college applications and SATs. Visiting my older sibling at the psych ward, I throw out the plastic knives before bringing in a bag of takeout spicy wings. To handle it all, I forge notes excusing my lateness to school and let all but a few friendships drop.

Nothing makes sense. I want out. I begin to dream of leaving home even if I can't shape a fantasy beyond that.

I save money, not knowing what I'll do, just that I must do something. I work in a small bead store and then at a fancy restaurant. Too young to drink, I can still serve alcohol. The restaurant has three floors, the lowest level with the Seekonk River. I balance trays of food through smoky rooms and carry buckets of dishes up outdoor stairs in the rain. The morning after my first shift, I feel like I've been beaten. Cuts from the dish bucket crisscross my palms and bruises bloom up and down my limbs. I get used to the heavy lifting and long hours. With each late night and double shift, I save my tips, dreaming, and asking *What if?*

When my friends go to college, I conduct my biggest experiment yet. I fly to Italy. It is 1998. I have a job as an au pair in the foothills of Mount Subasio in Umbria, the green heart of Italy. While admitted to college, I have deferred entrance and don't necessarily plan to enroll. My vice principal tells me I'll end up pregnant and on drugs if I don't go to college right away. But I have to believe the world is full of beauty. I see hints of wonder everywhere. I am ready to burst and waiting to happen. Neither my vice principal nor I has an inkling my adventure will lead to science.

I arrive to the smell of rosemary and sadness—the au pair family dad has a lover. The cold house sits isolated, far up a gravel road from Assisi. The mom and I fight over my insistence that the baby ride in a car seat when I drive her Fiat Panda, a roller skate of a car. She won't bother to teach me how to light the wood-burning stove and can no longer pay my stipend. The experiment fails.

I find an apartment in the nearby city of Perugia through friends from language class. In a city of young people from all over the world, we collide on the soccer fields, in nightclubs, in piazzas, in our shared apartments, and on *le Scalette,* the steps of the cathedral. In moments alone, I write—letters, journal entries, stories.

The year before I arrived, earthquakes had racked Assisi and Perugia. Some people still live in Red Cross trailers or thick canvas tents. Scaffolding embraces the Basilica di San Francesco d'Assisi. I don't know how or why the earth can ripple, liquefy.

Gathering with new friends from across Europe, Africa, the Middle East, and Australia, I see how our stories get shared to make each other laugh and cry. We cannot stop talking about where we come from. In every tale we swap, the landscape looms as a character.

A Swedish friend speaks of the booming sound of a river thawing in spring. An Icelandic friend tells of volcanoes erupting. Both talk about suicides during winter darkness. Friends describe holiday meals and summer dinner parties. These stories make me aware of *terroir* for the first time—we are of the soil and land.

My Fluff marshmallow spread, brought from the United States, and my stories of making out in parked cars amuse everyone—the built environment also a kind of terroir.

We think we have spun ourselves off from home, but I see how we reflect the mountains, rivers, trees, concrete, and disasters that fill our stories. Like Russian dolls, the stories fit inside us, and we fit into the landscapes from which we've come.

From overlooks, castle ruins, and mossy amphitheaters, I study the dramatic landscape of the Apennine Mountains—I want to know if each movement of rock and land has played out over millions of years or in sudden bursts, like the recent earth-quakes. *The hand of time passes on in music and in life.* I lust for words to talk about this.

I want to know how landscapes form. Someone tells me the deep pits of purple wildflowers on Mount Subasio are from mete-orite impacts. I wonder if we can date the punch of an extrater-restrial rock. I have no names for the plants or trees.

It strikes me that if I want to understand my new friends, or people in general, I need to know the larger story of the land on which their lives play out. The alchemy of my interest combines observation, story, and frustration at my ignorance of the land. This is my own logic to understand and make sense of the world.

A path abruptly appears in the murky, mostly blank, canvas of my future. My life experiences stack into cairns: invite me to pursue questions. I can leave the churn of childhood behind. I'm going to be a scientist.

Saying Yes to the Mountain

I

The airport lights flicker below, and Sig and I part in silence. I creep toward the women's cabin. Orange and pink bleed into my view of Juneau. The July sun has been setting since we snuck away from camp two hours ago. Sunset will run into the 3:00 a.m. sunrise. Camp will wake promptly at 7:30. I undress in the semi-dark, climb the damp wood rungs to my bunk, and listen for my seven sleeping female colleagues. I locate Kim's sleeping form, relieved I didn't wake her. The first day she grabbed my arm before I could worry about where to sit at breakfast, and we've been friends since.

This is field camp, a requirement for geology majors. Most of us graduate next year. The men's cabin is bigger, with twenty-some students. The fifteen staff sleep in makeshift bunks scattered throughout camp. We're a strange group, thrown together from across the country, with varying beliefs and ideas. Geology unites us. Everyone met ten days ago in Juneau, where we slept on the floor of the VFW hall, like a giant sleepover party.

A more typical field camp would focus on rocks and stratigraphy. But I'm interested in the liminal space between the slow boil of Earth's rocks and the atmosphere. This is the arena of life. I'm fascinated by the dance—the back-and-forth—between organisms and their environment, climate being a major tension. That the boundary conditions of my life could be ephemeral— bodies of water, diversity of species, and rivers of ice—puts me in awe. I describe myself as a plant ecologist who thinks on time-scales of thousands, or millions, of years.

For scientific purposes, glaciers are sensitive instruments recording climate. We sample the accumulation of snow over the last year, map melting along the edges, and measure the velocity of flowing ice. We follow in the footsteps of researchers who have come every summer since 1946. On that first expedition, the atmosphere had 310 parts per million of carbon dioxide. It is 2002, and the atmosphere now has more than 370 parts per million.

I know I should focus on the science, but I have a need dug deep by socks that won't dry, the cold of my synthetic sleeping bag, and the awkwardness of being in the wilderness with a group of strangers. Sig is on staff and ten years older. He introduced himself at the VFW by offering to help me pack, meaning help me accept all the things I could not bring along. We carry our belongings from camp to camp, traveling by foot or ski. Sig has shoulder-length blond hair and a sharp nose that divides his symmetrical face and blue eyes. I come up to his chest. The first time we kissed, he lifted me to his sun-chapped lips, and I wrapped my legs around his waist.

The program has a strict policy against romantic relationships. We are told not to massage each other's sore shoulders. The appearance of affection has no place on the ice. Sig and I avoid each other during the day.

Rain pounds on the metal roof. I pull my sleeping bag over my head and do ten crunches to warm myself. These mountains are the first place air hits resistance after leaving Japan. Heavy with water, the clouds release upon impact. In winter, snowflakes fall, never repeating the same pattern, but in summer it just rains or stagnates as fog.

2

We started in Juneau at dawn, hiked up through rainforest, reached timberline by lunch, and hiked onto the Lemon Creek Glacier in the early evening. An American flag, limp with rain, led us the last hour to Camp 27. We left behind a staircase kicked into the snow. Camp—a cluster of corrugated metal shacks anchored to raw bedrock with thick cables—greeted us with the promise of shelter and the smell of kerosene and mildew.

The ice field covers almost four thousand square kilometers. The ice beneath our skis is almost a kilometer and a half deep. Our path of about a hundred kilometers will take two months of zigzagging. Camps are one or two long days of skiing apart.

We students wonder which of us will come back as staff. The staff members were all once students, and, to us, that makes them the winners of past summers. Some are professors. One is entering medical school. One moved to Alaska to sleep in his car and chase the northern lights.

We set out a series of stakes in a straight line that runs from one side of the Lemon Creek Glacier to the other. We locate our geographic position at each stake with a GPS unit. After two weeks, the line of stakes becomes a V. Glaciers move under their own weight, pushed by gravity. They also move faster at the center, slower on the sides. The new geographic location at each stake gives us the distance the glacier has moved, and from this we calculate velocity. Information about the glacier's movements provides a window into the internal dynamics of the ice as it slides along its bed of rock and dirt.

On an expedition, we spend as much time maintaining structures, cleaning, staying warm, and cooking as we do conducting research. Part of conducting research is living in the wilderness.

Our leader, Maynard M. Miller, nicknamed M³—half scientist, half adventurer, a World War II veteran, and a member of the first American expedition to Everest—tells us, "If it weren't for the science, we'd be another lame Outward Bound program!"

3

We travel in small groups: eight students and two staff. The first hour of skiing to our second camp, I sing, alert and acutely aware of the dark peaks sticking out of 360 degrees of white. The next twelve hours fade into the timeless repetitive motion of kick-glide kick-glide.

We travel in a line, one behind the other. I am at the back. *I can do this. Why won't my legs stop burning? I can keep going.* If I go too slow, I'll lose the rhythm. Kick-glide kick-glide. Not fast enough to break a sweat. Damp air, or drizzle, and air temperatures in the fifties are perfect conditions for hypothermia.

It's late evening when my leg gives out. I fall onto the snowy slope. Arms and legs splayed, pinned down by my pack, I yell and yell in my head but can't get a word out. My knees stretch like rubber bands.

I get up and start again. The weight of my sixty-pound pack and my body rests on my thighs. Even though I keep the screams of hot pain in my head, my leg gives out again.

We had started at dawn. We have yet to eat dinner. I arrange my feet perpendicular to the slope. I use my poles to stand. The rest of the students are in a valley below. I see Jane, the other woman in our group. I tried talking to her last night—I asked if she was excited or nervous, but she shrugged and let the conversation die. She glides with momentum across the flat expanse, her thick brown braid flapping behind her.

I go ten feet and see where someone fell yesterday. The body imprint is active, smeared down the length of the slope. I don't know if the disrupted ski tracks throw off my balance or just my thoughts, but I fall. My body pressed into the ghost from the day before, I slide down.

Diego, the team leader, unhooks my skis, unclips my waist strap, and pulls my arms out of the shoulder straps. I sit up. He calls to the group below, "Set up camp!" We're an hour short of where we planned to stop. It is past 10:00 p.m. The moon is a perfect crescent over the tip of a peak, lined up as though it is balanced on the mountain. Diego pats my shoulder and tells me, "You cannot run away from things at home by coming to the ice field." He skis down the slope. In my exhaustion I think, *Speak for yourself.*

I stay put on the cold slope, comforting myself with color.

Tents pop up: yellow, red, green. This deep dusk-blue sky will last all night. The domes glow with the flashlights of students setting up sleeping bags. I wonder about Diego's comment. I had thought we were running toward something, not away. My teeth won't stop chattering. I wish my friend Kim were in my small travel group. I am relieved Sig is not.

When we are together, Kim and I braid our hair into pigtails and cover them with yellow or orange bandannas. This is her first time away from Oklahoma. I can't imagine that we will ever dismantle this snow globe, separate, go back home. That this community will come to an end.

Finally, I walk down the slope with my skis over my right shoulder, crashing through snow up to my thighs. I give up for the last bit and slide on my bottom, a ski in each hand.

It's only when I take my pack off that I find the cuts on my hips from the waist straps. In Juneau they told us, *On a sunny day, we are twenty minutes by helicopter to a hospital. But sometimes it's not flyable for weeks.* Dehydration! Hypothermia! Crippling blisters! Foot rot! The threat of the wilderness forces me present. This is nothing like science in a classroom. There are a thousand ways to die here. A thousand ways to fall into a thousand endless holes and cracks in the ice.

4

I wear gloves to write in my journal, but my fingers still go numb. What I love about wilderness is the same as what I love about stories. I can lose myself in it. I'm here to conduct science, but I need to metabolize the feel of a shovel in my hands and the sound of ice calving. Some moments it seems I have always known this story, I just hadn't yet seen the outline drawn black and clear by a steady hand. I jot lines of poetry at the back of my field notebook.

5

Danny leads fifteen of us armed with shovels. We aim for the X on a GPS screen, the exact site studied last year. Danny is a teddy bear of a middle-aged man with overgrown eyebrows and bushy ear hair who carries a kitchen's worth of spices in his pack. His leather boots are held together by duct tape. He has come to the ice field every other summer since he was a student twenty years ago.

Danny marks off a square of snow four meters by four meters. We dig. Every twenty minutes we switch off to rest, drink water, eat peanuts. Danny tells stories from past years, when he dug pits so deep they had to haul the snow out with a bucket-and-pulley system.

I chose my research focus, *mass balance*—the balance between how much snow accumulates and how much melts in a year—because I thought it would involve the most time outside. I wanted to see as much of the ice field as possible.

We dig until we find the surface of last year's snow. When we reach it—a boundary between white and white—we take samples from the pit walls. Someone will analyze these in a lab to calculate the volume of snow that fell over the past year. Brittle layers of ice tell of melt events, thick layers tell of storms. All together, the water content in the layers allows us to compare last year with previous years.

Eight hours into shoveling, countless times switching off, sweat drying, only for us to climb back down into the spiraling staircase of blinding snow, I ask, "Why don't we just take a core from the snow instead of digging?"

Danny shrugs. We follow in the footsteps of the past.

6

One year of low snowpack is an anecdote. But in the context of fifty years of data, we can show a wiggling, downward-sloping trend to the present. I am humbled to work for this year's data point. To collect bruises and strained muscles for what accumulates as one spark in the tail of a comet. We will package our summer into theses, add these to the greater collection of data, publish them for the world with the sharp focus of significant digits and calculated uncertainty. All this to demonstrate that our glaciers are melting: the globe is warming.

The numbers we collect on the ice field—the distance from the glaring surface of the glaciers to the bedrock below, the speed of mile-thick ice, the water equivalent of the year's snowfall—are incremental progress for science. But for me, they mean something more. I ski up mountains. Rappel into crevasses. Explore caves in the ice. The physical demands, the cooking and cleaning for fifty people without running water, the continual connection by radio—I didn't know I had the strength or capacity.

7

Fog wraps us tight.

The mass balance team is on a two-day trip. "It's the most beautiful place on the ice field!" Danny says. But it's twenty-five miles there, twenty-five miles back, and the fog never lifts: I never see more than ten feet in front of me. After we sing every song we know, a student yells, "Hey, can you ski like this?" He pulls his pants down to his knees. We ski bare-assed through the blinding whiteness, screaming at the top of our lungs.

Back at camp, Kim and I volunteer to paint the outhouse. The fog, like a blanket around the mountain, makes the air warm. It's hot in the outhouse. Paint drips on my pants and shirt. "I'm taking my clothes off." I throw my T-shirt and long underwear onto a rock outside. Kim strips down too, and we paint in underwear and sports bras. When you only have one set of clothing, not wearing it is your only available fashion statement.

The outhouses are set off from camp. They house emergency buttons, a last-ditch chance to call for help. If camp burns, the outhouse should be safe, but it's a long walk in the middle of the night. Announcements are made at breakfast reprimanding the men for peeing on the rocks in front of their big cabin. We handful of women pee on our rocks, too, but they don't smell. We don't have a critical mass.

"All my friends in Oklahoma are waiting to have sex till they are married," Kim says. "I'm the only one of my girlfriends who's not engaged. James and I are not waiting!" She declares, paintbrush held high.

"I don't know that I ever want to get married," I say.

"Really?! Do you want to be a professor?"

"You don't think you can get married and be an academic?" I pause. A puddle of paint collects on the floor. My father is an economics professor. My mother went back to school after her PhD in anthropology to get a social work degree. On weekends I'd watched my dad work through the glass doors of our dining room, his scuffed-up leather briefcase open and overflowing with papers, a yellow legal pad at the ready. He worked with urgent delight. I grew up hearing from him that other professions, things other parents did, were boring.

I let the silence linger a while before saying, "Did you hear about the outhouse that was here before? During a blizzard, right after this guy stepped out, a gust of wind came, and the outhouse blew away in chunks."

In this bare landscape, the four walls of an outhouse should stick out. Anytime we see strange shapes in the distance, I think, *Is that the outhouse?*

Kim says, "My twin! After we graduate in May, I will either move somewhere with my boyfriend or you."

Sig lives abroad and has brought up my coming to visit where he lives next summer. His invitation was more ambiguous. I couldn't tell if he meant, come because I'd have a free place to stay, or come because he wants me with him.

"Just think what we can do together," Kim says.

I'd scribbled a quote from another student on the ice in my

field notes: "Consciously stumbling, I plant my next step." I have no vision or plan for after graduation, but I am certain I need to figure it out by myself.

8

It's a helicopter day. Mail day! We get mail only when helicopters drop food, diesel, and equipment. I have yet to get my first cell phone, but, regardless, there is no Internet or cell service on the ice field. We relay messages to each other through walkie-talkies. I chew wads of tobacco another student had sent for, try not to gag, and then spit the brown ooze on the snow. Students give the helicopter pilot money for cigarettes he will deliver next time. I give him money for condoms.

With my small stack of mail, I walk away from camp to lie on a boulder warm from the sun. I slit open the envelopes and peer at familiar handwriting. I read out of order six postcards from my dad written over the course of three weeks. A postcard from a bar in New York with a penguin on it just says, "MWAH! Michelle." That's my favorite—just a kiss. Michelle is a friend I met the first day of college. I don't read the rest.

I peel off my clothes, which have not been washed for weeks. I rub snow over my skin like a bar of soap, and then close my eyes and turn to face the bright sky.

9

M³ tells us that science is about sacrifice: you follow ideas with no guarantee that they will work out. It is about pushing yourself to the limit physically and mentally, "but you also need shelter and warm food to do good science." He points proudly at the corrugated metal ceiling and plywood walls of the shack where we sit. On my last cooking duty, I start flipping pancakes before dawn and wash the last pots after nine o'clock in the evening. It takes a lot to maintain an expedition. We cycle through cooking shifts. I cross my fingers that tonight we will have canned peaches.

I don't know that I am good enough for this expedition. There were no fitness requirements. A month out, and I'm the only one who still falls skiing downhill. Sometimes I get too cold to think. My fingers can't hold a pencil. I'm recording measurements, but the numbers begin to take up half the page and then the entire page. I remind myself that we are here to push ourselves to the limit. I ski in the tracks of the person in front of me, in the tracks set fifty years ago by the first expedition on this ice field. When I feel too tired, I breathe in the eerie blue dusk reflected from all sides by snow and breathe out, "WAHOOOOO," at the top of my lungs.

10

Sig and I make love on a plank of wood over the entrance hall. Numerous makeshift beds in each camp, with makeshift ladders to climb, carry the marks of past expeditions. It is midmorning. We hear the door into the hall swing open. And close. The door into the kitchen swings open. Then closes. I'm on my belly, hips arched, Sig over me, one hand on my mouth. I smell kerosene from the lamps stored below. Through the curtain of Sig's hair, I watch the coils of rope hanging from the rafters sway as the doors below us open. And close. I listen to Sig's breath.

I saw a picture of lions making love once, the male over the female like this. She was reaching up, mouth open to roar, poised as if about to bite his neck. I feel the thick mess of my curls. Sig's hair is a knotted web I can't run my fingers through anymore. The chaos of wilderness is in our hair, our fingernails, our dirt-stained skin, our hearts. We strive to live up to the landscape.

11

My pinky won't bend. I wear duct tape on my heels to prevent blisters. I'm confused about the balance between risk and the value of data and experience. It's only one missed step in a jump across a crevasse that opens a hundred and fifty feet. In the vast expanse of white contours, our expedition is just a scattering of ants, dots on the landscape.

M[3] tell us, "We won't even try to retrieve your body if you fall into a moulin!"

We take turns going up close. One by one, we lie on our bellies and shimmy to the edge of the hole three feet across. Infinity stares back as a blue eye. Ice is a solid that can't make up its mind. It melts and refreezes. Water flows over and into the glacier. Trickles build to streams. A waterfall smashes through the moulin, and the ice shakes under us with the thunder.

I tell myself that I don't care what happens to my body after I die.

12

We ski down from a small rest stop on our way to Camp 18. We had to sweep mouse shit off the floor before putting our sleeping bags down last night. We leave early in the morning, the surface of the snow slick with ice carved into "sun cups," bowl-shaped inclines, one to three feet in diameter. I ski and fall. Ski and fall with nothing to dig my skis into. Topsy-turvy, I fall in and out of the sun cups. Everything looks the same right side up as upside down: white clouds and snow separated by a thin sliver of gray peaks. It's kick-glide then slide, knees bent, back straight, the weight of the world on my thighs.

Legs like jelly, I can't tell if the chatter in my teeth comes from the snow. Or if it is weakness, the fear getting me.

I think about my friend Michelle who sent the penguin postcard. She told me a story once about John Lennon and Yoko Ono falling in love. Before they knew each other, John went to one of Yoko's art shows. There was a ladder, and he could see a piece of paper at the top. He climbed the ladder and read the word, YES.

I think about saying yes to the mountain, relaxing my body, giving it over to the work of science. I'd written in the back of my field notes, "to think you are awake and then wake up again— push the limits farther and farther." I might fall, break my leg, crash through a crevasse. But I give myself to the mountain, skiing, falling, saying, Yes. Yes. Yes. Yes. Yes. Yes. Yes.

YES!

YES!

13

Our most scientific actions become myth as we perform the same procedures used in the late 1940s. The handle of my shovel is coarse with rust. I speak with the past in the repetition of bend, dig, lift, throw. I try to please the faceless voices I know by name and hometown from reading the ceilings of cookshacks. *Jeff Scottie, Denver, Colorado 1990 1991 1992 2002 . . .* Every camp is full of stories and song lyrics that chaotically fill the walls of buildings and the dark outhouses. *Beware the Center of the Llewellyn Glacier . . . What a long, long time to be gone and a short time to be here . . .* We try to catch up to the ghosts, to work as hard, to go back to the exact spots they went. The shovels stay the same, but the shovelers change. Every twenty minutes. Every summer.

When we find last year's surface of snow near Camp 18 and it is only three meters deep, I shudder: last year it was eight. I've fallen in love with a landscape only to witness it shrink toward probable, eventual, extinction.

I want to know the ache in my shoulders from shoveling twice as much snow, but my fingers and toes already burn with cold. All the past students who carved their names or song lyrics or jokes on walls, they all made it. *I'm one of them,* I tell myself.

On the ski back to camp, I hear the gurgle of an old-fashioned plane. Sig tells me, "That guy was a student in the program decades ago. He sometimes flies up here and drops ice cream or soda." He winks and skis off. I dream of mint chocolate chip ice cream. When I'm sick of the rain, I dream of snow.

I don't really know how far we are from civilization, but we joke that it is a six-day ski to the closest bar. When we finally step/slide off the last bit of ice, onto soft sand that doesn't come up to meet my foot but lets me sink in, I stare at the trees surrounding Atlin Lake. I can feel the smell of the trees on my skin—the presence of life knotted underground and reaching up into the sky. It's the freshness of oxygen in constant exchange between photosynthesis and respiration. Did they smell this too, the researchers from past years, as they crossed from ice to land?

14

Sig and I hike to a lookout spot of rock, the panorama of ice bright below. Sig pulls out a joint. I laugh. We make love on our knees in the open air.

We sit in silence for a long time staring at the yellow circles of lichen on rock, the grays and browns of dirty ice stretched to the horizon below, and the metal roofs of camp visible by the glint of sun.

"Where did you get weed?" I ask.

"I wanted to tell you, but I could get in a lot of trouble. I think you can handle being in on it now," Sig giggles. "I brought it from Juneau. I have been stoned ALL summer."

This man and I had an instant connection, and based on that, we trusted each other with our bodies. But he didn't trust me with weed? I study his hand, the skin thin and loose on the bone. I wonder if it is because I'm younger. I look at his big smile, a mouth so large it could swallow me whole. My heart and mind haven't caught up with each other. I might have thought this tryst a love affair, and the central story of the summer. But now it seems a distraction.

15

Restless, Jane skis in circles around us, "I just want to stay warm. I wish we didn't have to stop." It's in the forties or fifties, warm enough that the rest of us sit to drink water and eat mashed sandwiches. The occasional cloud makes shadows on the glacier. It is a beautiful day. I want to feel the sun, lie on my pack and rest.

When Diego calls for a break, Jane ignores him. She skis off, coming in and out of view on the rolling terrain. We continue at a pace calibrated to last for the long day ahead. Our breaks contract to a few sips of water, a handful of peanuts. Diego worries. No one talks or jokes. I sing Bob Marley under my breath, *rub it on my belly like guava jelly.* Eventually Diego speeds off to find Jane. The rest of us stay together, maintaining the steady pace. At least an hour passes before we catch up.

When I reach Jane, a knot twists inside me. My vision is funny, like I have red-colored glasses on. She sits between Diego's legs, limp and strange-looking. Something is wrong.

I empty my pack onto the snow: sweater, batteries, sleeping bag, dry socks, *A Pilgrim at Tinker Creek.* I undress and dress parts of her. Jane and I are not friends, but I am the other woman in our group. I swap her damp turtleneck for my sweater, her damp socks for a dry brittle pair of mine. Jane shivers, drenched in sweat.

The body can create its own natural disasters, small earthquakes that rupture. She is not okay. None of us know why—hypothermia, anxiety, not eating, appendicitis? We lay her on a tarp, wrap her in three sleeping bags.

"I'm so sorry. I was just so cold." Snot pours out her nose, her red face scrunched with the effort of tears. She repeats the same hysterical sentence, "I'm so sorry, guys. Guys, I'm sorry."

Diego radios camp, "We need help!"

I hold her head in my lap, pet her face like it's a small animal. "Think of home," I say.

"My parents are on vacation. No one is there," she sobs.

"Think of your favorite food."

"I can't take another day of canned green beans." Spit flies from her mouth.

Out of topics that feel organic, like mud or honey, I turn to her faith.

It's something she told me about a few weeks ago. Samantha, another student, and I were talking about gay communities on our campuses. When Samantha left for breakfast, I stayed behind.

"Excuse me," Jane said from behind me.

I jumped.

"I just want you to know that I am a Christian," she said, green bandanna over brown hair, face like stone. She stretched out *Christian* to emphasize her disapproval.

And here we are, two small dots thrown together, blinded by the reflection of the sun on ice a thousand years old. Her head in my lap, I drop the words *God* and *your faith* into her open mouth. She quiets.

The snowmobile arrives. Jane can't hold on to the driver. We stuff her into a basket on the back, and her arms and legs stick up out of the mesh.

I feel dirty, like we shared something empty, a quickie in an anonymous bathroom, a fuck in a parked car. I am ashamed at my easy out, using *her* God. Back at camp, Jane asks for me. I wait with her for the helicopter, help her board under swiping blades.

What I want to tell her is that we *are* making it, that we have to keep going, to see what the scientists who came before saw, to record this moment, add it to a collection of moments, like these, the last fifty years on the ice field, fifty other groups of explorers.

We made it. We are here.

Say yes. Just say yes.

So many small moments are what make a story, are what add up to infinity.

Give yourself over.

I only hope that what I feel, what I know in this moment, this truth, that I can remember it. That it won't fade like the patterns on a rock from the ocean when it dries. That when I'm down from the mountain, I will still know what I knew up high. Our bodies connect to the land, part of the web and wildness: a spirituality we all belong to simply by being alive.

16

We reach the névé line, where the snow stops and the blue ice starts. At the next camp, we will study streams that run over the ice, their volume, flow, and load another measure of glacial melt. We take off our skis and attach them to the sides of our packs. They stick up above our heads. The next camp is only three miles of walking or jumping across crevasses.

The ice is dirty, brown, and gritty. The August sun melts millions of tiny holes, so each footstep is a crunch-crunch, and I think of walking on the surface of a tooth riddled with cavities. I cannot stop turning to look back at the sublime white of snow and the outline of peaks I can navigate like a sailor looking at the stars. I feel empty despair, like I am walking away from a lover who is the whole world to me. Even if I come back, the ice field won't be the same. I keep singing. I don't have to cry because it's raining, and I've been wet all day, but I feel the distance of leaving the snow with each crunch-crunch, brittle ice breaking into jagged edges under my feet.

17

When I return to Providence for my senior year, I walk to class along Brooke Street. If a brook once ran here, the city buried it. The potholed pavement, the parking ramp, and the strip mall seal off any exchange with the soil. If I think about it too much, my lungs won't work.

I take a remote sensing class to process the data I collected and map the ice field's retreat over the past thirty years. The keyboard and slick computer screen are sterile, reflecting my still image sitting in a lab. I look through job postings, but nothing makes me want to run toward it. I date a beautiful and funny woman, but I'm newly aware of romance's fleeting wingbeat. Missing the physicality of the ice field, I do push-ups in the library stacks.

Most of my photographs from the summer are of mountains in shades of brown and gray with cracked or rippled snow and ice. The photos don't convey the vast connection I carry in my chest. I cover my walls with them anyway. I hang up a photo of Kim and me, naked but for boots, a picture taken to mark summiting a peak. Hair blows across our faces. We clutch each other, laughing so hard we'd fall if we let go.

One night, I take scissors to a picture of the ice field. I glue strips of the image down so the blank page alternates with images of mountains and ice. In these spaces, I find a rhythm for my breath.

Whatever Discomfort, Find Beauty

This is how it is at eleven thousand feet of elevation: you wake up to winter, eat lunch in summer, and put on a wool hat to sleep.

Cook oatmeal for the team at the picnic table. The hot coffee will stop your shivering. Just because you've been to happy hour with the other graduate student and the two lab staff, that doesn't mean you know them. Everything is different in the field.

By the time you arrive at Lily Lake, the Colorado sun will be above the trees and you'll be shedding layers.

The world is rapidly warming, even more so at altitude. It is 2005 and carbon dioxide concentrations in the atmosphere are 380 parts per million. You arm yourself with the theory of change: the past is key to anticipating the future.

Pulling cores of sediment out of Lily is like putting a finger over the top of a straw in a milkshake and pulling the straw out full of milky sweetness. You work from a wood platform strapped to two canoes. Push the core barrel down through the hole in the middle of the platform. That is your straw. The mud at the bottom of the lake is your milkshake. Each push starts easy, sliding through the water and loose mud, and then the slight resistance of water lily roots.

It is one thing to push, it is another to pull mud out. Your legs tangle in unbearable closeness with your team as you grunt and desperately give it all you have. Ignore the panic. It will come out. Remember to start low, use your legs and not your back. That everyone ten years older has herniated discs is a warning. There is no secret club to be admitted to, and if there were, it wouldn't be worth joining.

You lick your finger as each core comes up: *silt or clay or sand?* Evidence of times when the lake dried up and out here was sandy shoreline. You eat salami and cookies with muddy hands. If you had a mirror, you would see the streaks of sunscreen and dried mud on your face. There are clumps of mud in your hair and on your clothing.

The sun is brutal and hot in the thin air of the Never Summer Range. No matter how sweaty you are, keep your hat on. Splash yourself with the lake. Remember, if you swim, you swim in your clothing. Your bra will never dry.

Ignore the fear in your colleague's voice when he shouts, "Moose!" It is bigger than you imagined. Meaning, you didn't know how it would make your body feel. You cannot run. Your boat is anchored in place. So, take in the strong limbs, the reeds and water dripping from its mouth, the way the lake appears a puddle when a moose walks through.

This animal, it is part of the forest. But you? Your life is a blip, a yelp, a fleeting moment out of the thousands of years archived in the cores of mud piled up on the platform behind you.

When you need to go, don't wait. Don't stop drinking water, don't ignore the tightness in your throat, don't tell yourself you will ruin the team's groove. Ask everyone to turn around and sing *God Save the Queen.*

Squat over the hole. The moose doesn't care, the mud doesn't care. Look at the way light filters through pine needles, the outrageous lily flowers, the zing and zap of dragonflies.

God save the Queen!

The closer you get to the bottom, the more urgent it becomes to reach bottom. Bottom, as in the land carved out by glaciers in the last ice age that became the basin for water to collect. Keep everyone talking. Pause to eat chips or chocolate, whatever is at hand. The sun is falling quickly, the penetrating eye of a beast peering through the forest.

When the orange light turns the tree trunks to glowing embers, put long pants on over your shorts. Pull on your long-sleeved shirt. Take out your headlamp. If water or mud slides down the side of the core barrel and into your sleeve, let it warm you a moment before it chills you. When the air is still, listen for thrashing insect wings and jumping fish. This is why you are here. You insist, in your reasonable and rational way, on incessantly naming, measuring, and describing. All that just to participate, to be included, to be part of the great inhale and exhale.

Research as a Muse

The wind whips our small figures as we follow the trail of packed snow over the frozen waters of Medicine Lake in a suburb outside Minneapolis. I'm in graduate school, yet to pass my qualifying exams. Fishermen sit at random on upside-down buckets or inside structures with four walls, a roof, and lake ice for a floor. It takes us ten minutes to walk to the middle of the lake, pulling a sled loaded with scientific equipment. I had never seen people on a frozen lake before moving to Minnesota. My awe and nerves persist. We arrive in what looks like a village—the Art Shanty Projects—a winter arts festival. After we drop off the gear, I wander away. Our "performance" is not for a couple of hours.

Frozen lakes are anarchy at its best. Each winter, locals put up street signs and create roads, shacks/shanties/ice houses, and ice rinks. People gather for bonfires and pond hockey tournaments. They bike, snowshoe, walk, ski, skate, iceboat, and windsurf. Winter is a blank slate, and whatever happens on the lake is guaranteed to be wiped clean by the spring thaw. A self-organized Burning Man available on every lake, and, between Minnesota and Wisconsin, we have around thirty thousand. With my boot, I rub an iced-over fishing hole. Cracks radiate out from the basketball-size scar.

I open the door to an unmarked structure in the art village. A dance beat suctions me into a sweaty crowd. The eyes of the moving bodies peek out from between pillowy parkas and winter hats. Inside the installation, I dance. A bright-red scarf hides my smile.

Midday, we shovel away the snow in front of the Limnology Shanty, a makeshift museum devoted to the science of lakes put together by ecology graduate students. It's showtime. As geologists, we study lakes like detectives going through a suspect's trash. For us, a lake is a hole that things fall into, where, generally, they are preserved by the water. We read mud like a diary, transcribing it with equations, time series, and maps. All this is a story.

A crowd gathers around. The ice under my snow shovel reveals trapped air bubbles and folded leaves. We bore a hole with an auger, screw our coring equipment together, and lower a long metal apparatus into the dark water.

Bringing research to the public is not part of my graduate classwork, but it would be part of my job as a professor. Most funding for basic science comes from federal grants, and each grant must include "broader impacts." My PhD adviser claims training me, a woman, as one of his impacts. I think this is jumping the gun. His other PhD student dropped out of the program. We rarely hear the names of graduate students who leave academics. Asking questions, I begin to understand that there are many unnamed ghosts.

We extrude the core onto the snow. Visitors look down or kneel to take in the shifts from gray to brown to black mud. This is the story of ice sheets melting. Of conifers growing out of tundra. Of warming that shifted the prairie a hundred kilometers eastward thousands of years ago. Of the subsequent cooling and increased moisture that allowed the Big Woods of Laura Ingalls Wilder to grow.

We invite our audience to smear mud into the snow so they too will know that dirt records life.

With an audience—children, groups of friends, older couples, artists from other shanties—the work of science is no longer a means to an end. We are not just extracting a geologic archive for data points. The questions about ancient changes in the ecosystem, and what might happen in the future, sit unanswered, shared among us all, expert and layperson. All of us live as part of the landscape. And in our unknowing, with all our wonder and fear, we form community.

When I started graduate school, I decided that I could focus my creativity on statistical analysis and hypotheses about the climatic effects of grassland evolution. That would be enough. But other stories come to me as I underline findings in science papers and survey forests. They come as poetry, fiction, and essays. More than a diary. On a frozen lake, with an audience, in a shanty village, asking questions, what we have called science becomes art. I understand the flaw in my logic. Research itself is a muse.

II. Ecotones

1   a region of transition between two biological communities
      The *ecotone* between forest and prairie migrates eastward during dry climate conditions.

2   a time of transition between one phase of life and another
      We move back and forth across the *ecotone* between student and expert, as in a game of double Dutch.

3   a face-off with a choice
      Standing at the *ecotone,* you are torn between commitment to your research and the desire to work on environmental policy.

How to Pee Standing Up

◆

Rules for a Woman in Climate Science

Rule I. Don't Follow Your Heart

As a junior in college, I attend a Women in Science and Engineering (WISE) event and receive the first rule of working in male-dominated fields. Science professors sit around one corner of the table, students around another. Between us lies a gap of empty seats. We eat salad from greasy Tupperware or sandwiches from crinkling plastic bags. The student next to me vibrates the table with each crunch of her green apple.

"Ask us anything you want," a professor says with the sweep of an arm.

A student with a red thermos and black nail polish asks, "Did you feel like it was hard being a woman in science?"

The professors smile but don't answer. Finally, a woman in her late forties or early fifties leans forward, fingers pulling through her pixie cut. "I was often the only woman in graduate school classes. I am naturally quiet, but I forced myself to speak in class every day. Sometimes it took a lot to do that. It went against my personality." I understand what she is saying: *if there is no space for you, you must carve it out.*

In the 1970s, women made up only 10 percent of geology graduate students. Now, in the early aughts, that figure is over 30 percent. Regardless of the increase, the vast discrepancy between how many women are studying geology and the meager number in faculty positions is stark. I take this void as a promise of change, not a warning that women are allowed to go only so far. The imposter syndrome is in our heads. It's an internal problem, not the result of the bias of teachers, or mentors, or larger systems.

"How many of you have children?" I ask. I don't want children, or maybe I do. I ask as a metric.

After a long pause, a professor says, "Cheryl! Who isn't here. She has kids!" I think, *She isn't here because she has kids?*

A young woman with a brown bob says she is passionate about both social work and chemistry. "How do I decide?" She winces in anticipation of their response.

The professors look at each other as if they wish they could confer in private. One of them giggles. We all giggle. The professor who speaks doesn't sound nervous or uncertain. She says, "I don't think anyone here can tell you what to pursue. You should always follow your heart." She smiles at us.

I sigh. I hear everyone else sigh. This, to me, is an open permission slip: you make your own life. I flicker to another state of being. Life is not a test. Our hearts, bold and loud in our chests, are true. We can trust ourselves.

A knock jolts us, makes us turn. An environmental studies professor stands in the doorway. Even though we look at her, she knocks on the open door again. She doesn't smile when she sits in the open gap between students and professors. She spreads her legs wide and drapes her arms over the backs of empty chairs.

"No one tells a twenty-year-old man to follow his heart. Men are told to set goals. If you want to succeed, set goals." She looks around the room, not to make eye contact, but to make sure she has our attention.

I understand completely. Seize the day, carpe diem. Consume life in gulps. Go to extreme or remote field locations. Invent or build what you do not have. Run the most complex instruments. Stay on the cutting edge. And, if I put work ahead of everything else, I will be a scientist.

Rule 2. There Are Limits to What We Can Know

Thunder reverberates from the mountain dirt through the mini-van wheels and into our teeth. The car smells of dirt and sulfur. I look at Bryan, my PhD adviser, for reassurance, and he hands me the Nutella. I take a large scoop with a graham cracker. We couldn't find a knife, but as with everything on this expedition, we make do with what we have.

Our van sits in the Forest Service parking lot by Emerald Lake, outside Leadville, Colorado. I started my PhD program five months ago in Minnesota. For now, I collect lake sediment cores with Bryan. Where his research ends and mine begins is not yet clear.

On my first day of graduate school the program director asked if I had questions. No guidelines outlined the program's require-ments. The hope is to finish in five years, though I've met stu-dents pushing eight or nine. Another student in my department told me that when she asked her adviser about expectations, he said, "You have to fucking figure it out yourself." I told Bryan this, and he laughed. "It's kind of true," he said.

I asked the director about qualifying exams. His bushy eye-brows merged when he chuckled. He said, "We want to find your limits. We will ask you questions until we reach the end of your knowledge."

I look for role models, but while my department is large, there are only a handful of women on faculty. The youngest recently took a mental health leave from which she won't return.

Rain pelts the windshield, smearing our view. I look out at the impressionist brushstrokes of dark green, light green, and steel gray: spruce trees, lily pads, and the lake. Onshore, our work platform, a four-by-eight-foot board with a hole in the middle, is lashed on top of two canoes.

A bolt of lightning streaks the sky. "Ugh," Bryan says, but he grins.

"Bryan, we're collecting evidence of change in the water level of lakes, but what if those changes weren't from climate? What about the people who lived here thousands of years ago?" We want to know whether the water table fell, shrinking the lakes, when temperatures were warmer in the past. And if so, by how much? Knowing this can inform the design of climate models used to predict future droughts and wildfires.

Bryan uses a finger to get the last of the Nutella.

"Yeah. People could impact the land," he says.

"What about beavers?"

"Yeah."

"What about avalanches? Landslides!"

"Yeah," Bryan is still smiling, but he hunches his shoulders and wedges himself back and away into the corner of the car door.

"What about a stampede of elk?" I ask. We'd slept under the stars a few nights before, and I'd woken to elk grazing around our soggy sleeping bags.

"Anna," Bryan puts the Nutella on the dashboard. "You know, I don't know everything."

I look up at my teacher. I met him when I was an undergraduate in Rhode Island, moved to work with him, and now I schedule my life around his family. Sometimes I drive across town to meet with him at his house, where he spoon-feeds sweet potatoes to his baby while talking through my data analysis. Once, as I'm leaving, he lets me know his whole family has pink eye. He sets the rules. He is training me, teaching me to be a seeker of knowledge.

The rain is still falling, but the drops are smaller. Sunlight streams through the forest in long bright fingers.

"The Nutella is gone," Bryan says in a somber voice.

"Yeah," I say, but I'm listening to the *ping* of rain on the roof.

"Let's get back to work," he says.

I look out at the sun and rain. What keeps playing in my mind is the question of the limit of what we know versus the limits of what can be known. I wonder if our knowledge is limited by where our imagination falls off, as much as, or more than, it is by the boundaries of geologic preservation or technology. And what secrets does nature hold tight to, leaving no record for us to find and decipher?

Rule 3. Assimilate

With a wrench in each hand, I unscrew a three-meter metal barrel. "Green gray sediment with plant fossils. Oriented," I scribble. The drill tower's shadow falls across our workstation. All around our rig, the waters of Lake Bosumtwi glitter. We are in the middle of a lake about thirty kilometers from the city of Kumasi in Ghana.

Our expedition will run twenty-four hours a day, seven days a week, for two months. Each day costs tens of thousands of dollars, with funding provided by international governments. Our team is a mix of Ghanaian and American geologists.

I'd studied West African drumming in college, and I deferred starting my PhD program for the chance to come to Ghana as a field hand. When I ask the head American professor how he came to be in charge, he describes a scene like a dogfight. Whatever story is found in this mud will be published in scientific journals of the highest regard. I had to promise I would not jump the line for access to the samples. They will not be part of my PhD research. While my name won't be on the publications, my handwritten sediment descriptions will fill the official record researchers use for decades. And the analysis will take decades.

I watch the Ghanaian geologists greet each other with a handshake that pulls away like melted cheese and then breaks apart with a *snap!* I can tell which Americans have been here before: they shake-snap.

A million years ago, a meteorite impact created the lake basin. The impact breccia seals the lake off from groundwater, meaning the amount of water in the lake reflects the balance of rainfall and evaporation. Scrawny tree trunks with bare branches stick out of the shallows: water levels are currently rising. Recent flooding has forced families to move. Fish fossils in the hills above show that the lake was even higher in the past. Layers of sand in the sediment cores provide evidence of ancient periods when the lake was lower. Reconstructing changes in the water level of the lake can inform predictions of whether future warming will dry out or flood West Africa.

Twenty-four villages ring the lake. The steep crater walls, once forested, are cultivated with cassava, cocoa, and plantains, plants familiar to me from the greenhouse where I worked in college. My first instinct when I arrived in the village of Abono was to walk the perimeter of the lake. I've heard it takes three days. The lake is about ten kilometers across. But we only get time off if equipment breaks—and only for as long as it takes to fix it. At night, tiny lights shine along one arc of the lake. The electricity was part of an election promise never fully realized.

Local Ashanti oral tradition holds the lake sacred, home to ancestral spirits and the spirit of the lake. A long history with foreigners and mining companies raises suspicion about so much effort to collect mud. In observance of tradition, we sacrifice goats and chickens before putting the drill rig in the water. This too comes from our research budget.

Our team is split into two twelve-hour shifts. Each person shares a room with someone on the opposite schedule. At training, before the expedition, I asked who my roommate would be. "You are the only woman," one of the head professors told me.

"So, do I get my own room?"

"We'll see," he said.

The first day I am assigned to assist a pale, pear-shaped American graduate student. He's never worked on a rig before either.

"So, you are the only girl?" he asks. His tone borders between challenge and flirtation. "You know what everyone will be thinking after a few weeks?"

"About fucking you in the ass," I respond.

The men apologize for swearing in front of me, *pardon me, a lady is present, excuse me.* I measure, label, and cap plastic tubes of sediment, keeping the cores in order. *Fuck this, what the fuck, hand me that shit.* We ward off malaria by cursing. Ward off the spirits of the lake whose mud we take, meter by meter. I swear, and the men, my team, start to forget I am different. By the end of our first shift, I'm no longer an assistant. The twenty-five-by-sixty-foot metal rig is home.

The hotel maid sleeps on my bed during the day, the indent of her body pressed into the covers. My other roommates are lake sediment cores. The plastic tubes, each a meter and a half long, stink of sulfur and flake mud. My room offers additional air-conditioned storage. We have a shipping container lined with wooden shelves for the cores, but the container's electrical wiring sparked a fire, and the climate control is unreliable. Cores take over my balcony. Cores fill the space under my bed. I restack piles of cores to get to the bathroom. It is a room of my own with very little space for me. In total, we will collect more than two kilometers of sediment from the lake.

Amid the plastic tubes, I think through the choreography of the Ghanaian handshake. The feel of flesh held and then let go but not dropped. The continued tension as hands slide and fingers meet. The satisfaction of a thumb played off a thumb.

A boat drops us at the drill rig at six in the morning and takes the night shift back to our recently vacated hotel rooms. At six in the evening, this happens in reverse. When the boat docks, we walk one by one down a gangplank. Children call out and we toss candy. On my shoulder, I balance three one-and-a-half-meter-long plastic tubes the width of my upper arm. "Hey, lady!" a tall man with bright white teeth and high cheekbones calls out. "White woman, you are now a Black man," he says. He is not laughing: he is making a note. I look down. My pants, shirt, and steel-toed boots are covered in pipe dope, the black grease we use to lubricate threads on the rig. I scrub my skin until it is red and raw, but I can never get it all off.

With the feeling of a hard hat lingering on my temples, I go to the village bar. From shore, the rig looks like the eyes of a large creature in the lake. The bar is patched together with wood fencing, plywood, and metal scraps. Old men sit on benches. Children run through the crowd. I buy a Star beer and move into the heat of bodies. The young cook from the hotel dances with one knee up in the air. I sing along with the crowd, "one leg, one leg." The

Ghanaian highlife music spills out of the bar, and we dance out into the dirt road.

I dance with everyone and no one. Mostly, I dance with the throbbing night sky. I dance up to the cook. He puts his hand out, and I take hold. Life pulses back and forth between my body and his. Our hands slide against each other. I don't see how it happens because we are laughing. Snap! I want to ask to do it again, but he is already moving into the crowd.

Back at the hotel, I have only a few hours to sleep. The pads of my fingers trace my smile. Fingers that have touched the imprint of fish bones preserved in rock hundreds of thousands of years old. That have gripped metal tools. That have shot simple curses. And now, that vibrate with new language.

Rule 4. Safety First

We drive through the land ocean of North Park, Colorado—sagebrush with gray mountain dirt between scraggly bushes. Virga, fingers of rain that evaporate before touching the ground, play on the horizon. The road winds upward. Spruce and fir tower above as we pass through the ecotone above which trees can grow and below which it is too dry. The road ends. We must hike the rest of the way.

This is my one day with a team big enough to carry equipment above the tree line to American Lake, a key site for my PhD research. Two U.S. Forest Service pickup trucks pull up. Their open beds overflow with summer staff and a dog. I cold-called the rangers in the winter when I was spending long hours in a basement map library looking for possible field sites. After several conversations about terrain, trails, and the possibility of renting donkeys, they offered to come along.

My adviser, Bryan, arrives next. Last year he moved to a new position in Wyoming. He assumed I would follow. I didn't. Maybe it broke something between us. It definitely broke our lab—three rooms of computers, microscopes, and community. One of the undergraduates, the first in her family to go to college, left school. The other PhD student, a couple years ahead of me, dropped out. I see that brilliance has nothing to do with completion. Bryan runs a personal marathon of science, inspired and slightly manic. He's delighted for our company. But if I am going to finish, I am on my own. The practicalities of my education and career don't fall into the scope of his mad dash.

We unload the equipment from the van and onto our bodies. As we start up the mountain, the dog runs ahead and then loops back to our group of about a dozen.

"Should we be making noise so we don't surprise bears?" my labmate asks.

"You hear what they found in the bear scat?" a ranger says.

"What?"

"The hiker's bells." He laughs.

"Oh shit, not bear stories again," the other ranger says.

"Fuck that. It's smart to be afraid of bears," I say.

We walk on, weaving our groups together, laughing and cussing. I make a mental note to leave a bottle of whiskey at the ranger's office in thanks.

Forests don't end in hard straight lines. At high elevation, smaller and smaller groups of trees huddle together with more and more open tundra between. We reach a plateau with only a few krummholz, trees deformed and stunted by the wind. In the thin air, the sky is blue infused with blue. Nothing is between us and outer space. A valley stretches ahead with the waters of American Lake. The Elk Mountains sweep still higher around the lake, like the sides of a steel bathtub. It's July, but the winter ice on the lake broke up only last week. Patches of snow reflect the sun, making us squint.

We dump out our packs and pump up five truck-tire inner tubes. We lash the tubes together with rope and attach a thin plastic sheet with a hole in the middle. I look to Bryan for approval. He smiles, but his eyebrows form a V. He has not seen the platform I designed to be built on a limited budget and that could be hiked several miles. Before coming West, I talked to machinists, science elders, and coring experts. My field crew helped me test the platform on city lakes.

The ground around the lake is spongy, and my feet sploosh as I walk. White flowers with yellow centers and cabbage leaves grow in hummocks. What we want to know, the story written in the layers of mud beneath the lake, is whether trees grew higher in the past. We want to know if ancient periods of warmth, when the ice and snow melted earlier in the spring and later in the fall, allowed tree seedlings to grow in what today is tundra.

Bryan and I slip the platform into the water. A graduate student tows us to the middle of the lake, paddling an inflatable kayak. We anchor the wobbly mass of inflated tubes, not daring to stand up. I clip a pipe cutter and figure eight to the platform, so they won't float away—water comes up the center coring hole. We set the piston that will hold the sediment in place and lower the metal coring barrel into the water.

We add rods meter by meter, until we feel the lake bottom five and a half meters below. On our knees, counting to three, we push. The ice-free season is short at this elevation, and biological activity is much lower than in the forest. This should not take long.

"Ready?" Bryan asks.

"Ready," I say. Pulling the core out is always harder than pushing the casing in.

Kneeling, we clench our stomachs, shoulders, and legs against the suck of the lake bottom. Green water comes up over our knees. From here, the steel bathtub walls are anything but smooth. The slopes are covered in jagged rocks ranging from pebbles to boulders.

We have three cores of sediment when the yelling starts. It's clear there is more mud to core before we hit the basin's bottom. "What?" I shout. Onshore, our crew is small dots against the mountain. Bryan shrugs and we load the barrel for the next section. The kayak paddles toward us. "Lightning!" the paddler yells. I look up at the sky.

"What?" Bryan asks.

"Fuck!" I yell. "It's only ten a.m." The team I've amassed is not available tomorrow. This is it. What we collect today is the record we will have. And it's not just my research. All my cores will go into a repository available to the global scientific community.

"Lightning!" the paddler calls again.

"We have to finish this one," Bryan says. Anchored in the middle of a lake with our feet in the water, a metal rod in our hands, we are trapped. We're the tallest thing above tree line. I want to hurry, but there is no faster way to pull the equipment out of the mud.

"I thought we would be done by now. I've never seen a lake with this much sediment at such high elevation," I say.

Bryan grins. "It's awesome." The more sediment the better for resolving detail, but now we race time—and the probability of lightning.

"The rangers think you should get off," the paddler says.

A bolt of lightning strikes one peak over. I jerk. The platform lurches.

"It's sunny here," I whine. I want to stamp my foot, but I can't even stand. I look to Bryan.

"Up here lightning can switch from one peak to another without warning," he says.

"The folks onshore are nervous. They plan to start hiking down," the paddler says.

Bryan looks at me, "Well, Anna, it's your PhD. You decide."

I weigh a story about trees written in mud against our human lives. I'm at a boundary, my own personal ecotone between student and leader. "Let's keep going," I say, kneeling in the cool water with a metal rod held up to the violence of the sky. I would prove myself in this way too. No limits.

Rule 5. Literally, Not Figuratively

Ashanti mythology explains science through human experience. Respect for the spirits in Lake Bosumtwi sets practices in place that regulate and conserve fisheries and water quality. The lake is meromictic, the water in the basin doesn't mix, and almost no oxygen is present below ten meters. Below thirty meters, the water is completely anoxic, preserving annual layers of sediment that can be read like tree rings. In our cores, these layers look like the patterns in a tiger's-eye gemstone. On occasion, the spirits of the lake send a steady wind in the same direction long enough that water piles up at one end of the lake, and deep toxic water wells up at the other. In the myth, the spirits use gunpowder. The fish jump, gasping for air as hydrogen sulfide rises from the lake's depths. Fish float abundant on the surface, easy to harvest and good to eat.

I'm told fish kills happen once in seven years, or once a year, or only twice before today and in the 1970s. The science and the local information don't align. Their scales or reference points or purposes translate contradictory stories.

A rare event, and I am here to see it. But I need to pee.

The ship's portable toilet has been gone for several hours, taken to be emptied and cleaned onshore.

The men on the rig—as in, everyone but me—pee off the side. Not enough rain, and it starts to smell. For all our modern technology, our ability to see nanoparticles and explore outer space, we have yet to figure out fundamental and simple details like a system that ensures access to a bathroom. Of the four head professors running our expedition three are named John, one is named Chris, and none are women. It matters.

In the local village, work is divided by gender. Women collect water and firewood. Men fish. Women walk the hills with bright-colored cloths tied around their waists, babies on their backs, piles of sticks balanced on their heads, or branches across their shoulders with buckets of water on either side. Men catch fish straddling their *padua*, narrow slices of tree trunk paddled with bare hands or pieces of wood the size of playing cards.

The toilet on the ship had overflowed in the morning, sending a stream of stench running down the deck. Cleaning is women's work, but we hired through the town's "assembly man," something like a mayor, and he gave jobs only to men. The boat trip alone, to transport the toilet for emptying to and from shore, will take at least two hours. I have no idea when there will be a place for me to pee again.

When I was at geology field camp on the Juneau Ice Field, the eight women in the fifty-person expedition discussed how to pee on the open expanse of a glacier. *Do you wipe with snow? Stay on your skis!* An older student, normally reserved, blurted out, "If you just hold your labia back, you can pee a clean stream." I went over the word *labia* again and again. I'd never heard it said out loud before.

If I can't go to the bathroom, I'm not part of the team. I didn't anticipate something so mundane could slam a door on me.

I find a piece of plastic drill tube and wipe the mud off with my shirt. The tube is about ten inches long and three inches wide. I unzip my petroleum-stained khaki shorts, put the plastic tube in the fly, and pull my panties to the side. I hope the dying fish are more distracting than me: a woman surrounded by men, a white American in Ghana, a scientist ignorant of local gods.

In this moment, I set a new goal. I will pursue research I can do on my own—no logistics so expensive and complicated they involve professors in dogfights or reliance on the broader scientific culture. I don't want to be fitting myself into a woman's body that then has to fit into a man's world. I whisper, "Lay-be-a" to myself. The plastic is sharp, rough from being sawed off, but still, it is a relief.

Dear Spartina

◆

A Science Love Letter

Slow, sliding, smooth, shimmering, the river flows around my legs. I cross a flooded bridge and splash through the pink Maine dusk. The creek was dry this morning, but now my boots fill with water. In the marsh, tides mark time. I stumble, catch myself, the air cooling off and water wicking up my pants. All that keeps the ocean from reclaiming the land is the amalgamation of roots and shells that holds it together. Later, when the water returns to its cradle, the grasses will be left coated in white chunks of salt.

Do you know a word for the interior world shared by lovers? A place—a paracosm beyond others' imaginations. Sustaining love requires continually building that place, living in it, speaking its particular languages. It is a place to approach, but never arrive. Ambiguity, longing, and mystery fuel the approaching.

To love a landscape is no less effort, and no less imagined even in all its obvious palpability. My daily life as a field scientist is the repetitive prayer of muscle memory, the Latin naming system of plants, and the attempt to capture entropy by drawing it on a page. All this, a means to peer outside myself, to slow down, to look back, to wonder. To find wonder.

I learn the marsh by walking it: taste of salt, warm breeze, swollen mosquito bites on my forehead. I battle my way through waist-high grass and undulating mounds in the fresh tidal marsh to where I scan open vistas in the salty lagoons. The grasses leave linear cuts on my skin. The salty air curls my hair.

The river cuts the marsh with streams and creeks, dividing

the thick pungent mud alive with mussels, birds, and green vegetation. *Spartina* grass colonizes the land, locks the horizon in place, domesticates the churning mud and sand. *Spartina* possesses the magic of aerenchyma tissue that sucks oxygen out of the air and pumps it into the water-saturated ground. It is then that a neighborhood assembles: crabs scamper anxious to burrow around roots, young fish nestle safe in the flooded grass. *Spartina* is the continuity, the foundation of existence.

The marsh grass unveils in turn—seasonal cycles overlaid on daily tides. In the middle of July, *Puccenillia* begins to sing, dangling wheaty, soft, almost translucent flowers into the wind. In late August, it looks as if rice has been scattered over the lower marsh as *Spartina* flowers. In September, the color drains out from all the grasses except *Salicornia*. Stringy and segmented, it flushes a last wave of crimson.

I study modern settings where I can learn the landscape and use this language to reconstruct past ecosystems and climates. Episodic events like hurricanes show up in the marsh as thick layers of pebbles and sands. Droughts show up in lake sediments as sandy fingers where hundreds of years might be missing, winnowed away by lapping waves that moved inward as the lake shrank.

The repetitive prayer of muscle memory. Movements between lovers forge a bond that if broken would shred identity. In the same way, my body stores sequences of fieldwork and laboratory procedures.

Starting in Minnesota darkness, three of us drive north. By the time the sun is up the bare deciduous trees give way to the dark green of conifers.

The edge of the lake is a boundary with no meaning—the snow thigh deep, the lake ice two-feet thick beneath the snow. I place a rope across my chest and pull the loaded sled behind me. Between us, we pull thirty plastic tubes the width of my upper arm, fifteen metal rods the width of two fingers, measuring tape, sandwiches, chocolate bars, and a roll of duct tape.

We go to the middle of the lake, clear snow from the ice, auger a hole, and assemble our equipment. Five or six or ten hours pass with jokes and periods of silence, and we pull fifteen thousand years of mud from the lake. The time since ice sheets melted in this region.

When I return home from expeditions, I wake in the night thinking I'm in a tent. I free-fall in the stillness of my house, the aloneness of the indoors.

I spend months in the laboratory uncovering the one-and-a-half-meter sections of mud that I keep wrapped in layers of Saran, stored in plastic tubes, sorted on shelves in a walk-in refrigerator. I work through the length of them, sampling mud and extracting the lignin and waxes from fragments of wood and leaves. My hand squeezes the trigger of a spray bottle over and over. I sieve, search for a seed, a piece of cone, something with carbon that I can use to place the sequence of mud in time. A meter of sediment could contain thousands of years or just one catastrophic moment.

The eyepiece of the microscope barely touches my forehead, like a mother kissing a sleeping child. I draw the spiny monster forms of pollen in a notebook. For weeks, my world becomes a north–south grid viewed through the scope. Some grains are patterned like the fur of a tiger, others like a tortoise shell. I make notches in my notebook, 5 pine *(Pinus)*, 10 grass *(Poaceae)*, 20 ragweed *(Artemisia)*. Grass pollen are spheres with one hole, the edges raised like big puckering lips saying "O." Conifer pollen are different variations on Mickey Mouse, with a head and two big ears.

To go back to the lab in the middle of the night, I move the way I learned sneaking out of the house when I was in high school. I crawl down the stairs, skip the squeaky one. Bike in the dark, shivery, my body not fully awake. It's 2:30 a.m. I must smash together the five days I have on a machine to run samples. It's taken half a year of laboratory work to reach this point. The work has rhythm and little thought. I bike home for breakfast shaky from sleeplessness. I will do the same at dinner. My samples, ancient soils, tell of the evolution of grasslands and how it weaves into the story of horse evolution.

The Latin naming system of plants. To transcend mother tongues and connect scientists across the world there is a code with two parts. First the genus, a tribal name marking ancestry. Second, the species, a specific name that describes simple features, *Quercus rubus* a red oak, *Quercus alba* a white oak.

For me, the language of botany transformed the green brushstroke of landscape into distinct stories. Mint has leaves that grow in pairs opposite each other, but at ninety-degree angles from the leaf-set below, *opposite decussate.* When I learned this term, I saw something new—opposite decussate roadside weeds. This feature, combined with a square stem and the lips of the flower, gives mint away. I find it in the Rocky Mountain wilderness, a first clue of past disturbance. Only hours later do I find the ruins of a long-ago mining camp.

I go to the public library in Walden, Colorado, to spend an afternoon in the men's bathroom. Just steps from the urinal is a collection of pressed and cataloged plants stacked in a metal cabinet. When adolescent boys and men in cowboy hats come in, I step into the hallway. I scribble plant names and draw leaves and flowers. I'm voracious. Without names, I don't know for sure if what I saw yesterday is what I saw today.

I learn to identify plants by small clues, a piece of a leaf or needle, a grain of pollen, chemical fossils measured in machines. I wake in the night to churn through ideas. I jot notes in a small journal I always keep in my bag. My mind lingers on details. I puzzle. I dream.

I read the landscape. Each plant is a story with a range of conditions it can live in—soil types, temperature extremes, drought tolerance, salt tolerance. The collection of plants holds meaning, tells of the land and climate. When I braid these together, they build a rich narrative of ice ages, drought, the formation of ecosystems. But these puzzles are missing too many pieces—nothing is certain. The distance this creates leaves something like the longing of separated lovers.

Capturing entropy on the page. All the scientific process, the encyclopedic knowledge, the photographs of mountains and oceans—these mean nothing until translated. The translation of the relentless and indifferent actions of nature is made with equations, diagrams, statistics, graphs. All the dirt and color and wind removed to leave only two dimensions. My art is to structure that love letter, to turn my rambles through the woods and marsh, my protocols in the laboratory, and my pollen counts into patterns that form a narrative. To be able to share the story I read in the land.

I sit alone at my desk with my field notes from the summer, the pages smudged with dirt or chocolate:

> July 5th Hike to Lake Agnus, 3153 meters
> Area looks like may have been logged in the past.
> Off side of trail still some snow patches.
> Visible moose droppings. Forest is pretty dense.

During our courtship, my husband and I lived apart for months at a time. We wrote love letters. He sent a whole page describing me walking across a room, the way he would touch my shoulder first. The details of an old house told of the bigger fantasy we were building—the sun through leaded windows, the smell of bread baking. He conjured an image we both held tight.

He once sent half a loaf of bread by express mail so we could share the same breakfast. I had just started graduate school and lived in a large co-op house with a mound of shoes by the door. I watched from the sidelines as the mound grew, dispersed, and was stacked high again with the coming and going of my five housemates, overnight guests, friends joining for dinner, late-night band practices, and concerts that made the floors bounce as though the wood was elastic.

Alone in my office it is like this: no kiss of the microscope on my face, no razor grass edges, no murmur of mosquitoes, no depth of mud holding me at a frozen lake after dark. I look at my computer screen and I think of the weeks I lived without electricity. I remember how my body fit into the landscape just so, how I washed myself by swimming in lakes with floating ice. My office furniture is not ergonomic, does not hold me or make me feel alive.

I take the raw data from running samples of mud through instruments in the laboratory, the data ordered by depth, depth converted to time. I can't smell the trees or feel the sun, but thousands of years of time flow through my mind and I see the retreating ice sheets, growing and shrinking lakes, the march of deciduous trees northward. I'm no longer held to my dusty university office, not pinned to the present moment. This is time unobserved, the lover baking bread to send in the mail, hands kneading the dough, yeast brewing and pushing the wet mound up and up.

From Scientist to Animal

The Trinity Site is a fenced-in area of desert about the size of a football field outside Albuquerque, New Mexico. The first atomic bomb was detonated here in the summer of 1945. The same open flat land, covered by scraggly tall weeds, lone tufts of grass, and bare gray dirt, lies inside and outside the fence. I'm four months pregnant, with the end of my PhD in sight. It hadn't occurred to me that a conference on grassland ecology would take me to a nuclear bomb test site. I'm not showing yet, and no one knows I'm pregnant. I signed a waiver acknowledging the risk of stepping on undetonated explosives, but it didn't mention anything about radiation.

A monument of black lava marks the location of the explosion crater. The government bulldozed the green glass that formed from liquefied sand and bits of bomb during detonation. Black-and-white photographs along the fence document assembly of the bomb. Descriptions on placards accompany the photographs, as if this were a museum and not metal chain links that separate desert from desert. The photographs show limber young men with full heads of hair. They strap equipment onto a sedan to drive to the site.

They were young scientists, like me, probably enthusiastic to explore how the world works. I am angry that this too is science. I walk slowly toward the gate. It's 97 degrees Fahrenheit, full August sun. *I don't want to faint here.* I concede their tremendous accomplishment amid my horror. When the test bomb went off here, people in Japan still had twenty-one days to go about their daily lives. Three weeks to not know that a nuclear bomb can photograph a person's shadow onto the sidewalk.

Standing still, I consider that my research builds upon fifty years of systematic rainwater monitoring for nuclear test fallout. I use these data to set up a baseline for tracking markers of rain, snow, and evaporation in lake water through Earth's history. From this I reconstruct how past changes in climate drove the migration of trees and animals. And yet. My tools were developed for military interests. I want nothing to do with destruction or war.

A sign tells me an hour inside the fenced area will expose me to half a millirem of radiation. Hand to belly, I think of Byrd. We named the fetus Byrd on the spring morning we found out I was pregnant. We had gone outside to examine our bewilderment in the early morning light. The perfect body of a cardinal lay on the sidewalk in front of our apartment. Though five weeks pregnant, I was convinced that the moment the test stick showed two pink lines, a piece of the bird's soul dropped down into my womb. We will give him a proper name when he is born, but only then.

The sign continues with information about common ways we receive radiation in our daily lives. My body, the vessel for this baby, cannot protect against gamma rays. On a cross-country flight a person receives about two millirem of radiation. I want to leave this place. I contemplate renting a car to drive home. I imagine the conversation with my husband, Dan. He would weigh the risk of driving two thousand miles against the risk of two millirem of radiation.

I walk out the gate, back into plain desert. I expect the exquisite heat to let up, but it's the same. I kneel to inspect the vertebrae of an animal on the ground—a piece of jaw attached to spine. The other scientists on the field trip move in groups from photograph to photograph. Stair-step limestone mountains frame this small piece of earth. The fenced-in plot of land is obliterated when I step back and see the horizon of mountains reaching beyond.

I sit alone on the bus. I think up excuses to use later for not drinking beer. I think about how I have avoided telling colleagues that I'm pregnant. *I must tell Bryan,* my PhD adviser. No matter

the event—a wedding, an exam, a canoe trip, a work conference—
my pregnant body and this fetus are the subplot. My mind for-
ages for fresh food, scouts opportunities to sit. I don't feel like
chatting about research with strangers while the pregnancy plot
flows through my mind.

But I also have to figure out my future. I have a fellowship
this year. Byrd will be born halfway through, and I will defend
my PhD by the time it ends. I don't know what happens after
that. This conference is full of researchers I admire and potential
leads. As we drive away from the Trinity Site, I watch the expanse
of desert, ocean-like, out the bus window. I feel lonesome, like a
child sitting alone in an elementary school cafeteria. I keep my
hand over my belly to hold Byrd the best I can until we pass out
of the gate.

In my early twenties I sought spirituality through change: dat-
ing someone new, moving, leaving a job. It was like turning over
rocks to find secret worlds on the other side. In graduate school I
race, throwing myself ahead. I get to my office before eight in the
morning, stay until hunger drives me home at night. I burn out
each academic year with classwork, lab analysis, paper writing,
and teaching.

In the summers, I go west to collect samples of mud and water
from lakes. I survey trees and wildflowers in mountain forest and
tundra. Sleeping outside, away from computers, phones, electricity, and running water, I reincarnate. My body, as much as my
mind, becomes a tool for work, and inside me tension and bitterness unravel. I return home strong and happy. In the mountains I know the purpose of life. Back home I remember the *feeling* of being part of something larger, but I can't maintain that
connection to the dramatically textured landscape. On campus,
science is discovery through procedures in the lab or data analysis. The spirituality of this work is in the intensity of the pace and
the complexity of the questions.

A female professor in my department tells graduate students,
"My husband and I thought that if we worked hard enough, for

long enough, something would come through." They lived airplane rides apart for several years before finding jobs in the same place. Then they had one child. I don't question why such rational and intelligent people would convince themselves of an academic rapture. I too dream of a tenure-track position.

One day I realized the idea of children was in my heart. The seed of motherhood already rooted. I felt a palpable emptiness. Dan and I talked round and round. We couldn't rationally decide to have a baby. It would mean not living apart, not following the best jobs, and then perhaps not making it. We had no backup career plan. Our bodies decided for us. We found each other again and again, unable to choose with our minds, we answered with limbs and mouths, calling out each other's names.

When I tell my dad I'm pregnant, his joy comes first. But it's only moments later that he asks, "What about your research?" I'd submitted a proposal to the National Science Foundation for a postdoctoral fellowship. Over the past year and a half, I had met with potential advisers and laid out hypotheses. If I get the fellowship, I will have resources for both my family and my research, as in, freedom.

I am exhausted yet insomniac, hungry yet repulsed by food, glad that my body works yet cramped. A friend tells me how she loved her babies from the first moment she knew she was pregnant, how she loved never feeling alone. For me, I need to be alone to become aware of myself. In winter, I walk out to the middle of frozen lakes to find absolute quiet. I don't feel a presence with me as I deliver my talk in Albuquerque, as I fly home, as I go to my office, as I make love to Dan.

I hear the heartbeat at the midwife's office and cry at the fury of effort, the speeding thump-thump. On my bike commute I imagine the heartbeat in my pelvis and the heartbeat in my chest. I imagine a double glow from each spot. But I don't use the word *love*. I still can't believe I can grow a baby. It's beautiful—amazing—miraculous. But I feel too unfinished and uncertain to actually be a mother.

The first kicks are like a small animal roaming inside me.

Sometimes Byrd feels like one of the invasive eighty-pound carp in the Mississippi you catch with a garbage pail lid for self-defense. The fetus is closer to the size of a goldfish. My initial reaction, from the gut, is to make the movement stop.

I call the nurse line and burst into tears. "I'm craving blood," I say, embarrassed and afraid. Biking to campus, my legs pump the pedals round and round, and I hold the image of a sweating glass of red liquid. In my office I try to work, but my mind returns to the glass, the deep color. I dream at night in the bold red, orange, and yellow brushstrokes of a cave painting. Animals with manes blow in the wind. I am only lines of hair and limbs, my pregnant belly an exaggerated globe in orbit.

My pica craving for blood is in the category of dirt, rocks, rust, laundry detergent, burned matches, and chalk. The nurse sighs, "You are probably anemic. Just take iron supplements." She couldn't be more bored.

In the sixth month Byrd wakes me with frantic movements at night, like a swarm of bees let loose. I stop shaving, figure nature is taking me back. My internal organs shift into new configurations as my uterus expands. When I touch Dan's lean body I'm surprised, as if I expected the whole world to inflate with me, as if we are all in a Botero painting. I no longer fit into normal clothing. I read in a pregnancy book how I might feel ridiculous having sex, "like a rhinoceros." But I know I'm ripe, bursting with excruciating, awesome, and profound potential. I wash my hair with vinegar and baking soda. Tufts of armpit hair are now a part of my beautiful animal body working.

At seven months the movements don't wake me. The kicks are less sudden. Byrd now does swim-bys—like a sea mammal swimming toward a crowd in an aquarium, pressing its body against the glass. I wonder if Byrd is trying to get out, to make contact, or just turning. A friend tells me how she didn't realize until she was twelve years old that her mother didn't know everything she was thinking. I'll be observing this child forever and I will never know Byrd's mind.

I ask my stepmother when I will become a mother. She says, "When the baby is born." The risk of miscarriage needles me. Pregnancy books remind me the baby could die at seven months, eight months, at any moment. I read that I should consider my options in case the baby dies in childbirth. I discard the idea that I am a little bit more of a mother the longer I carry Byrd. A former roommate told me that after her abortion, she knew she was a mother, and the soul would come back to her in the future. I try to imagine Byrd's soul waiting for me all these years, but what comes to mind are filing cabinets, suitcases, and aquarium tanks.

At the pool I wear a black-and-white polka-dot bikini, my eight-month pregnant belly bare but for the lightning strikes of stretch marks. An old woman in the shower sings, "She wore an itsy-bitsy, teeny-weeny, yellow polka-dot bikini." I don't like the publicness of my pregnant body. As a scientist, I mostly exist as a mind. Drafting my dissertation, I remove myself as the protagonist. *Water was sampled from the 100 lakes in July of 2006.* No grit or sweat included. The language of science disembodies the researcher, making scientists interchangeable, the work reproducible. But a pregnant body needs—water, food, a place to rest. All my need is on display.

My body forced onto me, I can no longer narrow the world and disappear completely into research projects. An engineering professor told a group of students that when she was pregnant, she vowed to her department head that nothing would change. I feel the pressure to stay the same. But I am already different.

"Is it a boy or a girl?" the woman in the shower asks.

"We didn't find out," I say. It felt too intrusive watching Byrd in the ultrasound, seeing the mouth open and close, the flow of blood, the bones through skin.

In the pool with Dan, one of Byrd's limbs presses out below my ribs, and I wiggle it back in. I like that our family of three are all in the water, I feel closer to Byrd. *We are all in this together,* I say with arms and legs moving through the water.

At a Christmas party, voices burst from the house. Shoes of all sizes barricade the entry. There's an eddy of children in costume, children in red jumpers, children entangled in cat's cradle. I am eight and a half months pregnant and too large to squeeze past a group talking on the stairs. "Excuse me," I say to a thin man in a plaid shirt. He smiles and steps back an inch. It's not enough. I smile. He walks up three stairs and onto the landing. Growing up as one of seven kids in two households, each merged with another family, sometimes volatile, I learned to wedge myself into nooks. There was no space for me to take up. As groups move to let me up the stairs, my belly huge and vibrating, I feel a wicked sense of glee.

In each bedroom a circle of people play music. In one room they sing pirate songs, in the next, banjo, guitar, mandolin, and fiddle radiate out as new players join at the edges. Despite the cacophony of music, I'm sleepy, legs like tree trunks, feet flattened by the extra weight.

The children at the party make the miracle and strangeness of pregnancy normal. And that normalcy carries great relief. The day before, a graduate student said to me, "You know that it is basically a parasite?" And then she asked if she could touch my belly. I hear researchers in my lab speculate about a past graduate student who got pregnant and dropped out. Studies discover over and over that motherhood is a major reason women leave science, leak out of the pipeline. A professor in my department tells me that she went back to teaching the day after giving birth. One of my advisers tells me she took five days off. *It is all you need,* she said, looking me in the eye. I want to ask the parents at this party if pregnancy also forced them to bridge their minds and bodies. I want to know if it changed their needs.

I realize that neither cookies nor music nor splashing water on my face will make me less sleepy. I open the door to the room where we threw our coats. Six teenage girls lying on the bed roll their eyes. They cuddle closer together, hands clasped, bodies pressed like spoons. They look like vampires waiting for sundown with their black eyeliner smudged from rooting around or tears or just not being washed off for days. One girl whines, "You bumped me." She sits up, moves her body away from the pile. Another girl comes to check on her. She straddles her, sitting on her chest, and lays her friend back down. I remember this intimacy with my high school girlfriends. I wonder at it now, nonsexual physical intimacy. I find my coat in a pile by the window and go out, closing the door behind me.

I don't know that it is my final week of pregnancy. My due date is two weeks off, but with a wide margin of error. I don't know that these are my last days to jump in a car without thought. My last days to be alone. On Monday I hear from the National Science Foundation that I will not get a postdoctoral fellowship. That leaves me four months to find a job before my current fellowship and health insurance run out.

I can't wear clothes on Tuesday, just a blanket wrapped around my waist, the pressing, pushing, fizzing in my belly too much. From the four-season porch of our apartment, with the winter sun on my back, I email and talk with other potential postdoctoral advisers. We spin out ideas, discuss upcoming deadlines. I start my days at seven and work until Dan comes home in the evening. I don't want to make love. I don't swim to break up work. I need to finish more applications. I am like a puma on the hunt— patient, silent, untiring, focused. Before bed, I reread my work from the day. Then Dan sits behind me, and we breathe together. I try to match his slow rhythmic breath, unravel enough to sleep.

Thursday night our neighbors come for dinner and we dance in the kitchen. After they leave, Dan gets an email congratulating him on receiving a postdoctoral fellowship at Penn State. I'm terrified. My target has narrowed to one institution.

Friday morning, I wake to wetness between my legs. An unscented colorless fluid forms a puddle when I stand. I have received a bad lottery ticket. The general hospital rule is that a baby should be born within twenty-four hours of water breaking. And for a first-time mom whose water breaks before labor starts (about 25 percent), it will take twenty-four hours to go into labor.

I reread my applications one more time, go to the post office with a cloth diaper in my underwear, and then call my midwife.

In the hospital, I put my foot on the shower stool and lunge. Water pounds me. I rub my nipples, turn my face to the forceful stream. A clock is ticking. I rub my clitoris. The contractions come as if I am a windup doll, my nipples and clitoris the crank. I must keep labor going. These places of deep intimacy, of sexual allure, transform in the effort to open the tight purse string of my cervix that holds Byrd inside.

The contractions reel me. My water broke twenty-five hours ago, and labor only started in the past few hours. I grasp the bar on the wall of the shower. *The pain is with me, the pain is part of me,* I chant. *Yes, yes,* I say out loud. I focus on opening. *There is only the present. The past and future are not real.* I press into a wall or grab Dan as I am rolled and thrown. I'm in a realm I've never known. If I were ever to speak in tongues, it would be now. If I were to lead a séance, in these moments I would be able. I walk the halls of the hospital in an emerald-green cotton dress. The nurses urge me to trade it for a hospital gown. They don't want me to ruin my clothing, but I am my own goddess. I follow the intensity.

At thirty hours, a petite nurse with corkscrew curls puts a hand on my arm, "We are going to start you on Pitocin." I feel as if she slapped my face. Synthetic hormones will intensify contractions and induce labor. Half the shock of (my planned) pregnancy has been embodiment. I don't want technology to make this more efficient.

It is one thing to surf *my* body's pain. Another to be throttled by synthetic pain. They hook me to an IV. I stand across from the nurse, the bed between us. "Anna, now is the time to go into the pain," she says. All I can do is trust her.

I look but cannot make sense of what I see. I shake. My teeth chatter. They lay Byrd on my chest. I count four toes. Nurses stand on either side of me, push with their whole body weight into my abdomen. I scream. I count four toes, again. The nurses press with their hands. I scream. The nurses pause.

"It's totally okay, but he only has four toes," I whisper to Dan. "There are five toes." Dan pets my face.

I feel Byrd wriggle on my chest, wet with amniotic fluid and blood. I count, again, four toes again. The nurses press harder into me. All my control—the intentional breath—is gone. I scream and scream, nothing left to give, and the nurses push into my weary body.

"You are hemorrhaging," they tell me, perhaps for the second or third time. This might be the story of my death in a past century. I feel movement and wonder what's on me. I remember Byrd. "Someone keep their hand on the baby, I can't hold on to him," I call out.

"I've got him, I'm right here," Dan says.

Unbelievable, a bit of Dan, a bit of me, and then my body grew a fetus. My womb opened and a baby was born. Every person was made this way, I remind myself when I hear about a murderer on the news, when students file into my classroom in sweatpants smelling of hotdogs. My amazement seems a measure of the distance between myself and nature. My human hubris that we are tame. That we are beyond wildness.

As we leave the hospital a nurse reassures me, "Go home and lay on the couch. Take your shirt off, just lay the baby on your chest. That's all he wants. He could pick your shirt out from a pile of shirts. He knew your smell before he was born." All I want is to inhabit the world of smell and touch and breath. But when I sleep, I dream of blood raining down. I read that inducing labor increases the risk of hemorrhaging.

My father, a childhood friend, a stranger ask, "Was it natural?"

"I did not use pain medications, but . . ."

"Good," they say before I finish. I wonder about "natural" being reserved for the experience of pain. Like pain was the final ingredient to make me a mother.

For the first week at home, I can't stand and hold the baby without feeling faint. Dan brings me pot roast for breakfast, I am ravenous and anemic.

"You didn't use pain medication?" other friends and strangers exclaim. "You know you don't get a gold star for that." Every birth story is heard as a fable, whether it wants to be or not. When people ask for my fable, or just the juicy bits, panic tingles up my back.

We are cautious initiating our delicate routines with Byrd. The first bath at home is more complex than any procedure in the laboratory. Our neighbor comes over and unswaddles him. She waves his thin arms and legs in the air, giggles and blows on him. She laughs. I don't know this side of motherhood. I am raw and in love. I see we are missing part of the point, the joy.

Three weeks after the birth, we go out as a family to fulfill our need to eat colors and ideas beyond the cycle of feeding, diapers, and sleep. Between bright paintings on the walls of the Walker Art Museum, we find a black curtain into a dark room. I sink onto one of the four carpeted benches in the middle of the room, still weak. On a screen as large as in a movie theater, images play of black-and-white clouds. Orchestra music blares. An atomic blast goes off at Bikini Atoll. Waves radiate out toward a dense scattering of battleships.

I think again of the great team who developed *the bomb*. The great curiosity that led to great destructive power. That too is science. The process of asking questions and finding ways to answer them is beautiful. But discoveries can serve a wide range of purposes: military, forensic, extractive, medical, business, and even poetic.

The suspense, risk, and innovation of my work have given me purpose, and that purpose allowed me to live in my head. But now Byrd needs my body. Could I learn to forget myself again? Would I want to?

Byrd is warm and sleepy from being carried on Dan's chest. I lick his forehead to wake him, blow on the wet spot. My bare breast reflects the light in the film: the light of sky and clouds and sun and water. His mouth suckles the air. He turns his head left and then right, searching. I position him and he opens his mouth as if to bite into a hamburger. When I kiss Dan now I cannot help but imagine that we are suckling each other, a muscle memory of comfort. I thought I'd found breathing skin for the first time kissing as a teen. I thought I was discovering a new force. Now I use my whole body to hold Byrd, feed him, sing to him.

The pediatrician's office had called because I scored poorly on a screening. The multiple-choice questions on postpartum depression surveys reminded me of teen magazine quizzes. My midwife sat next to me and asked me to tell her what was wrong. Flora, Fauna, or Merryweather, I couldn't decide which fairy godmother she was. She did not reduce my experience to a problem in my head.

"I can't tell if the baby is breathing," I said. Sometimes I lean so far into the crib I start to fall in and have to catch myself. Dan tells me not to worry. Then I have to worry for the both of us. I am keeping alive a fragile new creature with a body that has been split open, bled, and sewn back together. My recovery happens while I'm making and giving sustenance to Byrd. I know nothing about babies or caring for them. Natural instinct can't teach me how to operate a breast pump.

"If his face is pink, he is breathing," the midwife said. This helps. She gives me something concrete in an ocean of overwhelm.

The bomb goes off over and over, we see it from above, from one side and then another. I wonder about the sea creatures in the waters below, the reef, the sunken skeleton ships. Dan and I will move forward in time as parents and as scientists: the questions infinite. The movie is silent for a period of explosions and then the blasts screech and echo. Byrd's eyes open for a moment, then close. My nipples tingle. I will be a mother with my whole body forever. Steam and water move up, making me think of a flower that keeps blooming. There is no undoing what has been done. Deep in my breast tingles. The water moves outward in a ripple of waves that engulf ship after ship. I want to build us a home that holds us like a fox's den.

An Interview with the FBI

It's a kind of poetry to collect vials of mud and grass by canoe. A kind of poetry to burn the hell out of things, smash and shatter them. I identify plant molecules in ancient sediments by how they break apart, measure their stable isotope ratios for carbon, nitrogen, oxygen, and hydrogen. This is how I read the landscape. This is how I reconstruct ancient periods of climate change.

But I need the grit for daily living: health insurance, a steady income, a place to set down roots with my baby. I'm looking for a future. I move from Minnesota to Pennsylvania for a postdoctoral fellowship at the end of 2010 when Byrd is eleven months old. My fellowship will last only two years.

Being a scientist is a path made by crashing through the forest at dusk. In geology, the science experiments happened thousands or millions of years ago, and I am looking for puzzle pieces from those ancient events. With enough lines of evidence, I'm able to reconstruct shifts in the boundary between forest and prairie, and further back in time, the evolution of grasslands.

I never have all the pieces to complete the picture. Discoveries only lead to more questions. Research is making my map as I go. This is how new knowledge is forged.

As a postdoc, I use forklifts to move gas tanks, troubleshoot plumbing and electrical wiring of mass spectrometers, and pull all-nighters in a windowless room with roaring Frankenstein instruments held together by salvaged parts. Colleagues in the oil industry tell me how university laboratories are outdated, how we are working ten years in the past. Fracking technology has initiated a gas boom. Oil companies are hiring. The chemical fossils

I use to research ancient climate change can also characterize oil deposits and guide drilling. My multivariate statistical analysis on conference posters draws fossil fuel industry leaders. If I come for a job interview, they will fly my whole family out with me. *But I am a climate scientist!*

My mouth waters at what industry can offer. The newest equipment, lab technicians, family leave policies, regular work hours, paid vacations. I want all that.

My postdoc adviser and boss, and the tenured professor to whom my lab belongs, mentors me with warnings that my ideas are too big and too new for a woman. She tells me that even if I get a job as a professor, there aren't enough grants to go around. Being a professor is my dream, but that dream now depends on her letter of recommendation. She won't write one until she knows me well enough—this she says will take at least a year. She forbids my working on prior research, limiting my ability to publish, the prime currency for academic jobs. In short, I belong to her.

The students in the lab come to me fretful and crying. I am all that stands between them and the professor. She breezes into the lab to use the microwave or electric kettle. Technically, brewing tea a few feet from someone using hydrofluoric acid is against safety standards. But it's not just the culture here, the machismo of a geochemistry lab is no less than on a drill rig. Chemicals are made safe telling anyone with concerns to fuck off.

My dream of being a professor is not limited to excitement for research questions. I also want a platform to communicate climate science to the public. I ask Professor Richard Alley to lunch to talk about how I can connect my research to present-day climate policy. I want to participate.

In Alley's PBS special *Earth: The Operators' Manual,* he introduces himself as a Republican who plays soccer on Saturday and goes to church on Sunday. I see him running to and from campus with a backpack holding what he needs for the day.

Alley and I sit in a window booth with a view of students hurrying to class. He asks for his salad dressing on the side and to hold the mayo on his sandwich. I tell him that I, too, am passionate about communicating science to the public. He does not smile. He asks if I know that I cannot do any of that now. There can be only one plotline. Until I get tenure, I must focus on publishing my research. He is kind. I know his advice is good: anything to keep the dream alive. But after lunch I can't get it out of my head that staying in research means a life without mayo.

It's not just science communication that lights me up. Since Byrd's birth, I've been writing essays and stories, some built from my notes taken in the field. Anton Chekhov once said, "Medicine is my lawful wife and literature my mistress." I write early in the morning or in the middle of the night—whenever I can. It goes without saying that I do not talk about *this* writing in my postdoc lab.

Writing up my research for peer-reviewed journals, I scrub out the hours of labor, the meals cooked outside, the moving of campsites as wildfires encroach, the shock of jumping into a lake with icebergs, the wounds sustained, the sunburns, the personal jokes, the frustrations, and the bonds formed. While I erase myself from scientific discoveries, I can bring my whole self into creative writing.

I am awarded a yearlong fellowship to study writing craft with a cohort in Minnesota. Once I start my postdoc, I fly back and stay with friends for weekend workshops. I pump milk and store the bags out the window in snow on the roof. Professional writers give lectures and assignments to teach us about storytelling. To create a narrative, something has to be at stake. The character must change. If the writing brings disquiet, it calls for further exploration. *Go toward the discomfort.*

Having been asked to consider changing my story from climate science researcher to fossil fuel extractor, I look for other possible threads. *I want to make a difference.* The molecular fossils I use to reconstruct ancient periods of climate change are also used for forensics. Geographic provenance can be traced to map the migration of fish, verify the claims of fancy champagne, and solve murder cases. Perhaps I could contribute in this way. When an unexpected job pops up on one of the listservs I anxiously scan to start my day, I jump.

At the checkpoint for the military base with the FBI laboratory, I recoil, an involuntary reaction to armed men. I don't relax as I enter. Visiting the Federal Bureau of Investigation laboratories in Virginia, I will find out whether, instead of unraveling the intricacies of Earth's climate system, I could do forensics on bombs. My postdoc ends in a handful of months, maybe, depending on whether my adviser wants it extended. She is holding that funding over my head.

By working for the FBI, I could slip my adviser's knot without leaving science. In the same way I use chemical signatures to reconstruct ancient climates, I could use them to determine the manufacturing processes of explosives or the geographic regions where the cotton used for wicks was grown.

Inside the FBI building, I walk through several labs. The sound of an explosion outside makes me jump. The woman showing me her laboratory doesn't react.

"What are they doing?" I ask.

"Practicing raid and hostage situations," she says. Next to the building is a life-size diorama of a small town. The sound of bullets hitting the town diner, the sidewalk on Main Street, and the cute theater—even if part of a practice—makes me squirm. The blasts don't shake the building, but I still feel the world tilt as I consider the need to prepare for the destruction of people, homes, and the wildness of trees and birds.

At lunch I sit with a large group. "Have you been to the Body Farm?" a woman asks.

I shake my head and take a bite of my grilled cheese sandwich—a Kraft single on white bread.

"People donate their bodies after they die to the Farm," the woman says.

"Some bodies are hung, some are left on the ground, and some are buried in shallow graves," a woman with a big smile jumps in.

"What do you study at the Farm?" I ask to be polite.

"We study what it is that dogs smell when they search for dead bodies."

I smile and nod at the table of friendly people who are learning an alphabet to read molecules of rot. I tell myself that this could be my everyday, these scientists, these sciencey conversations, this cafeteria food. An explosion booms outside the window.

When I feel the familiar glow, the sense of grace, I am standing in front of an audience of FBI employees to give my research talk: on water taken up by blades of grass that carry signatures of rain or snow, of plenty or of drought. My audience doesn't just ask about the utility of my research for forensics, they ask all kinds of questions. I've taken them into the beauty of Earth's ancient forests and plains. In this room, with my data and hypotheses and these curious scientists, I transcend my concerns about what it will take to secure an academic job, what it would mean to have one, and what it would mean *to walk away.*

I'd recently flown back to Minnesota to give a literary reading. From the stage I looked into the faces of so many friends and so many strangers. Never mind the swirling in my gut, I commanded the room. Tendrils connected me to every person seated in the auditorium with its exposed brick walls and scavenger hunt of Greek letters pressed into the floor. I had thought writing my story was sorcery. But reading aloud, I learned another kind of magic. The audience's anticipation eddied in my mouth. The characters from my pen flew over the crowd. That bright shining alive feeling is also in me as I give my talk to FBI employees and answer their questions.

There are a lot of stories and headlines about women leaving science. If I leave, I will fulfill the stereotype: woman scientist has kid = woman scientist leaves science.

Geoscience undergraduate students are now evenly divided by gender, but at each level in the hierarchy the field skews more and more male. It's not just that I have to prove myself as a woman, the anemia is broader. Geosciences are the least ethnically and racially diverse of any scientific fields.

There are different ways of knowing and different ways of asking. What questions might a Native American researcher have? What new angle might a Latinx researcher take? What doors might be opened by those who grew up blind or with limited mobility? What knowledge are we missing by supporting research led by a narrow slice of humanity?

While I am white, whiteness was not inevitable at my public schools in Providence, Rhode Island. Black and brown were the colors of our prom court and class presidents. Brown University was only a mile from my high school, but I felt like I'd taken a wrong turn when I got there. During orientation the university's president told us we were the best and the brightest. *Where was everybody?* I could see that opportunities didn't align with talent: the university wasn't need-blind in 1999.

"How do we know her ideas are her ideas?" a professor said in a faculty meeting I attended during graduate school, dismissing the one woman of color interviewed for a tenure-track job. In my nonvoting and nonspeaking role as a student representative, I waited, an ant squashed under a boot, for someone to defend the researcher's record as her own. While I was in the department, every job search resulted in the hire of a man.

If academic research is not offering the cutting edge in humanity or equipment or support, the beauty of molecules might not be enough.

After my talk, I tour FBI facilities where evidence is processed. Each workstation has a high table with a roll of brown paper. I watch a woman with a calm demeanor neatly wrap a lump of crime scene.

We walk down a hallway with many framed photographs. I stop in front of a stylish and artsy picture. I can't tell what I'm looking at. A building falling? A placard reads, "Bali, 2002 explosion, 202 people killed."

All down the hall are bright and glossy images of life torn apart in an instant. These are a kind of art, but one that reminds me of the potential we each hold to be wretched. That is a curiosity, a form of exploration, a question to be asked.

"The Firearm Museum!" my guide points out. "Patty Hearst's gun! Baby Face Nelson's gun!" Dark wood and metal, a wall of triggers and barrels. Death and power and fear. Collected together, the weapons, objects of crime, lose their individual identities. They are like a rainbow of beetles pinned in a display that no longer gives the feeling of insects crawling on the back of the neck. In the museum, the guns are beautiful.

In a large room where the walls are lined with what look like dishwashers, a man asks, "You got a dollar bill?" He wears a white lab coat over jeans.

I pull out my wallet.

The man puts the bill under a green laser. "Ha! Cocaine. I knew it."

"I don't—" I start, my disorientation makes the room wiggle.

"Every bill has cocaine on it," he laughs and hands me back the dollar.

I follow him around the room as he talks, "That instrument was just for analyzing money. We calibrate our instruments for particular substances so that they are always ready. That one over there is for lubricants, that one for skin, that one for nails, that one for soil."

I look around, breathe in the sterile room with its geometric lines and murmuring engines. In the invisible web that pulls at us all, connects us through the air we breathe and the ocean breeze that tugs at our hair, it is questions that lead to possibility. That is the race in science. To read what we do not know. This is why I have come. To have access to facilities that would allow me to ask questions.

I understand that holding steady to a well-defined path is not a story. The road divides, branching—fractal. I don't want to be the mom who leaves science. Science defines my identity, but the needle is stalled and the thread tangling. I need to go toward the discomfort. I understand that something more than a professional title is at stake.

I have to let my own poetry unfold—even if it breaks my heart to walk away from my research. This is crashing through the forest at dusk, making my map as I go. Shallow graves and guns can be poetry, but that is not my kind of poetry.

III. On the Hill

My postdoctoral adviser tells me that if I leave research to work in Congress, I can never come back. The scientist in me looks in the mirror at the mother in me. "Let's go," they say at the same time. Caring for an infant, I understand that the passing of time means no do-overs. As my baby changes and grows, seemingly overnight, our environmental crisis intensifies. The future is not distant. Immersed in a new vocabulary of footie, binkie, and crying it out, shards of clarity crystalize. Ice sheets melt. Species go extinct. Wildfires and hurricanes come more frequently. Why are we not taking action? I have to find out how the sausage is made.

View from the Lactation Room at the White House

We make fast friends—we're both lost in front of our destination. The Eisenhower Executive Office Building, housing the executive offices of the White House, takes up four city blocks. It's so large we don't know where to enter. It looks like a cross between a castle and a spaceship, an architecture style called Second Empire.

"Women's Energy Summit?" I ask, taking a guess.

We fall into step together, both anxious to not be late.

In D.C., I talk to strangers everywhere: to find out why a train is delayed, because I like their scarf, without reason. This is a city of transients and compulsive networkers. People are as open to meeting as if we were all on a study abroad program. A colleague tells me that every conversation should be treated as a job interview, but for me it's curiosity—I talk to strangers because I can.

My friend for the day, Jessie, is elegant in all the ways I am not. Her suit has a metallic sheen, it fits perfectly over her thin frame. She is shy, smiles flicker across her face and disappear. Runner's legs float inches off the ground on snakeskin heels. Her perfection is made human by wild red curls moving in all directions.

My hair is similarly wild, but in contrast to her feminine sleek, I still look pregnant. My suit is too tight on the top and too big on the bottom—my breasts are oversize canteens of milk. I wear orthopedic black flats. My shoulders tilt, the unequal weight of a purse with notebook and business cards on one side, and on the other a weekend travel bag with breast pump and paraphernalia smashed in. Senator Franken hired me six months pregnant as a fellow on his energy, environment, and agriculture portfolio. Before taking maternity leave, I had just three months of briefing the senator, taking meetings, and researching policies. I've been back in the office for only six weeks.

I treat every conference like the apocalypse has happened and I'm forging a new life. Whether it's a day or a week, I build a social network as if the event will last forever and I need a posse to survive. It's like I can't conceive of the ephemeral nature of folding chairs and scheduled keynotes.

Like Dorothy, I have three key characters in my postapocalyptic conference world. First, a confidante, the person to sneak out with and go sit in the sun. Second, a Madonna/mother figure, someone who will save me a seat and look out for me (and vice versa). Third, a homeland connection, someone with a tie to the outside world who offers a reality check. All this falls under networking, but my goal is just to feel comfortable.

For this one-day event, my confidante will be Joe, a twenty-four-year-old man interning at the White House who I was told would have the key to the lactation room. My Madonna, Jessie with the red hair, will be the one to bring me into conversations with her large smile. My homeland connection is another scientist from my fellowship program. While I'll never see the first two again, with the last I will share a Passover seder and swap emails over the next several years as we navigate ourselves out of D.C. and into more permanent lives.

The invitation from the White House is a beacon of hope, a signal that I'm going in the right direction. Up until eight months ago, I was on the path of a research scientist. I came to Congress through an American Association for the Advancement of Science (AAAS) fellowship program that brings scientists into government.

In science, dressing up was wearing yoga or camping clothes—pretty much anything from REI. I never owned a purse and felt like a child who'd stolen her mother's makeup when I tried to paint my face or nails. In D.C., everyone looks professional. This is the only place I've ever been where it doesn't undermine women to look sexy—it doesn't decrease how seriously people take them. These are the powerful, and no one is fucking around.

In contrast to the sterile, slow, and silent isolation of science labs, policy is fast, a jazzy intellectual challenge of intensity and improvisation. I'm armed only with my background in botany and chemistry, the ability to read geologic maps, and my personality. I learn in the first week to run in heels, keep a poker face, and politely listen without committing to idealistic and impossible pleas.

While everyone in D.C., on any side of any aisle, talks about protecting American families, many don't have children. D.C. is young and workaholic. People have kids in their late thirties, forties, fifties, or later. I often take Byrd, my three-year-old, to play with a friend at the park while his friend's dad and I hang out. His friend's dad is sixty-five.

Being pregnant in D.C. is another jump into the unknown. People ask, "You're only thirty-two, and this is your second child?" And then the follow-up: "Are you religious?" Another fellow expresses her concern: "If you don't go to happy hours, how will you ever get jobs?" I don't bother telling her that at toddler birthday parties we also exchange business cards.

My fellowship program provided a month of training, a salary, and an evening mixer with congressional offices to kick off the interview process. Unleashed, entering the mixer, none of us certain what constituted professional attire, I had the fast-talking, big-smile feeling I imagine one has rushing a sorority. Except I was rushing a sorority six months pregnant.

In an interview with the lead office of the bipartisan climate and energy bill, I brought up being pregnant. I thought my interviewer, a male in his late twenties, might not have noticed. The spark of connection went out with him saying, "Yeah, well, you know, come January, I'm really gonna need someone up and running." I had asked my AAAS program officer whether a dress with a suit jacket was equivalent to a suit. She said, "If you don't have any clothes that fit you, just do your best." I was advised to work for Guam or another nonvoting territory. Instead, I went full steam ahead. I didn't bring up pregnancy again. These moments of negativity didn't foreshadow what was to come for me.

I got offers, lots of offers. In many interviews, and with Senator Franken's office, I met a new breed, those who believe working mothers are the same as everyone else. I waver on this. I'm shocked each time I slam into a physical boundary—pain, fatigue, hunger. Hunger without warning sometimes throws me off a cliff and leaves me shaky and confused. But these are obstacles that require acceptance and planning, not things that should prevent my being hired. Only I know what I'm up for and what is too much—and the answer to that is different for every woman.

The red-haired Madonna and I find the security line to enter the White House. "Well, my husband and John Kerry are very close," she says. We've both worked in Africa on climate change, she in development, me collecting geologic samples. My fieldwork experiences always offer some reliable connection to others. When I meet with lobbyists and other interest groups in Franken's office, it creates intimacy to ask where in Minnesota they live or grew up, and then name a lake I have surveyed or cored near their town.

The beep-beep of a text message interrupts our conversation. I peek at my phone: "Just wanted to give you a heads-up, day care has put up a sign saying that several kids in Byrd's room have lice." I shove the phone into my purse and feel a flush of heat across the back of my neck. Dan will be home with the baby, Cluck, for the next three months. Byrd still goes to day care, living at a pace too fast for the sleepy quiet days of a new baby. I immediately feel itchiness on my scalp and imagine lice jumping on my curls like a trampoline.

We used "Cluck" as an in utero name, riffing off a T-shirt I once gave Dan. It says, "Sometimes love has a chicken's face." I bought it at the Wild Rumpus bookstore in Minneapolis, where cats and chickens roam freely among the bookshelves. My idea of a utopia.

We show our IDs to an armed man in uniform in a small guardhouse. "You cannot leave now that you're in," he explains.

"Your clearance is just for the one entry." He sends us through a turnstile, and we wait on marble steps.

My phone beeps again. "You forgot a piece of your pump, should I bring it to you?" My stomach twists and I squeeze my toes and fists to remind myself to stay present.

We go through another security check with metal detectors. Uniformed officers open our bags. Of all the possible evils the scanner can detect, none can expose that I may be bringing lice into the White House.

A hundred women pack into ten rows of wooden chairs so tight our shoulders and legs touch. The room has several windows with white molding above and thick maroon blackout curtains. I imagine the lice, no bigger than sesame seeds, jumping from head to head. I consider that it might be reasonable to not want new mothers around. I cry more easily, I leak milk and smell like yogurt, I leave early to pick up my kids, and, now, I carry lice.

My husband texts, "Got pump parts, I'm outside. Guard says I can't go in. Light rain, but able to keep Cluck dry." It's 10:30 a.m., and my breasts are filled with wet concrete. "I can't leave and come back in," I text. I need the intern with the key, and I need my pump part.

Community activists are on the agenda at 11:45 a.m., and I must pump before they speak. My breasts are a new frontier of emotion. Hot dampness catches me by surprise. My shirt soaked through by the news of a tornado in Oklahoma. Tragedy that before would have felt like a dagger to the heart now actually shoots pain from collarbone to nipple, leaving me breathless.

The new secretary of energy, Ernest Moniz, speaks to us. It's his third day on the job. He's had a fresh haircut since I saw him at his confirmation hearing. I step out just as an astronaut is about to speak. She is a leader at the National Oceanic and Atmospheric Administration (NOAA), where I would love to work. I find Joe, the intern with the key. His smile is a smirk, but he doesn't have the key. He has to go get it. I show him a picture of Dan and explain about the forgotten pump part. He skips down the stairs promising to be back.

"Hand off complete," my husband texts five minutes later.

I wait for Joe to reappear.

I talk farm bill in the hall with a lanky young man, also interning at the White House, also a male staffing this celebration of women in energy policy. We talk USDA. I ask where the lactation room key is kept, how long Joe might be. He shrugs.

This is the first time I've had to talk to a man about breastfeeding, other than family or a medical provider. In the Senate offices, I've never had to explain my lactation needs. Other people get coffee, walk dogs, talk to their brother on the phone, and they don't sit their supervisor down to explain themselves. It's assumed that they, like me, take natural breaks in an intense workday. And the Senate is set up to accommodate pumping. A system is in place, so I'm not asking for what feels like a favor on a case-by-case basis. The Senate accommodates not only staff but also the public. A nursing mom can lobby her senator and then pump milk in a clean private room. The system assumes participation by lactating moms.

It took me months before I realized how many moms work in the Senate. I only heard about the Senate Mom Group from a chance encounter. The meetings are advertised by word of mouth. The first time I went, we gathered in the conference room of a senator from the extreme other end of the aisle from Franken. But for this hour, politics didn't matter. Like fish gasping for air, we talked quickly. Women jumped from composure to tears without hesitation. I said little. I was overwhelmed by the fact that I wasn't alone. I recognized colleagues in the room who I hadn't realized were also moms. They talked about a husband with cancer, being in the groove of balancing everything, in-law complaints, in-law praise, taking cookware to a restaurant and paying for Thanksgiving sides to be packaged as if they were homemade.

Twenty more minutes go by. I peek in on the summit. I come back into the hall. Sweat soaks through my suit coat, and my breasts are harder than the marble walls. I'm still shocked at their size, my body as new and unknown as during puberty.

Joe hasn't returned. It's been an hour. I watch the end of the panel from the doorway, one foot in the room, one in the hallway.

I try to imagine where in the ten acres of this office building the magical mothers' room might be. And unlike the lactation room in the Senate, I've been told, it is just an empty room: no sink to wash hands, no fridge to keep milk cold, no hospital-grade pump to save lugging my own equipment around.

Finally, Joe. He's walking down the hall, smoothie in hand. "I can take you now."

I follow him along the black-and-white diamond floor tiles. We skirt around a large crowd of men in dark suits. The ceilings are seventeen feet high, with light fixtures that belong in a Victorian dining room. The hallway is long, with white walls and evenly spaced doors. White pillars stand guard on either side of each door.

We take a turn, and another turn, and come back to where we started. Joe is starting to sweat too. He tosses his cup into a trash can. The building has two miles of hallways. This could take a while.

We go down a different hall. "I think we go up these stairs," he says. He leads me up a spiral snail shell. Above the stairs is a dome with a twenty-foot-long stained-glass skylight of red, white, and mostly blue. Ornate gold-plated molding slides along the curve from skylight to wall. The natural light is bright and joyful, but I am shaky with the poison of milk sitting too long. Milk held in is like unexpressed anger. It can make you sick. I've had mastitis that started with the random thought of throwing up, and fifteen minutes later I was on the floor unable to stand. Breasts aren't designed as long-term holding tanks. Especially in the early months following childbirth.

"Wait here!" Joe says, and he runs down the hall, a flash of black suit.

In a breastfeeding class someone asked where to pump if your job requires driving. "Church parking lots," the teacher said. "They are a sanctuary, after all."

He runs back, "Found it."

I follow him through a small door and into a secret gnome hallway. While the floors in the main part of the building span seven black diamonds elbow to elbow, here they span just two. If I reach my arms out, I can touch the walls and the ceiling. The narrow galley bends and turns every ten to twenty feet.

Joe stops in front of a closed door. "Do you want me to wait outside?"

I feel the discomfort of a woman alone with an unfamiliar male in an alleyway. This isn't a part of the building filled with people making the country run. This is where dusty things are stored.

"No thank you." I slam the door, unable to ask the obvious. It seems ridiculous to staff a conference celebrating women with men—to task a young man with keeping the key to the lactation room.

Finally, my sanctuary. The critical link that connects all my education and work experience to all the potential I have to offer the world. In these early days of motherhood, lactation rooms are what keep me from falling off the sheer drop on either side of a steep path.

If I don't pump during the day, my flow will dry up. While there are health benefits to breast milk, babies who drink formula also grow up bright, brilliant, and beloved. My commitment to breastfeed is based on the physical experience of a river of life flowing through me. To give that up would be a kind of spiritual death. Not rational, purely animal.

I enter the room so relieved I could barf. I might actually need to barf. But I don't think these thoughts at that moment. Instead, I'm struck by the view.

The view from the lactation room at the White House is of the Washington Monument and the Thomas Jefferson Memorial, the Tidal Basin, and God's clouds dark and fast across the blue sky.

The view is framed by just a small attic window, no decorative molding, a semicircle above a square of glass. There is indeed a fridge and a hospital-grade pump, and I understand that the conference organizers never bothered to check the room when I emailed ahead with questions.

But the curtains are open, and from up here D.C. is ribbons of forest weaving around expanses of prairie. Aside from the parked cars, it's a timeless view. Nothing of the overcrowded streets and traffic. I push the window open, and in glides a cool breeze.

I throw my bag down on the gold-brown wall-to-wall carpet and lay out my supplies on a small table. I attach plastic flanges to bottles, tubes to pump, pump to electrical outlet. The setup of the pump recalls the muscle memory of setting up a science experiment in the laboratory. The results come with similar satisfaction, two or three or four ounces, measurable results. The pump runs, milk flows, and I close my eyes. I feel the sun and breeze, and, finally, I relax.

Lactation rooms are the rare places where I form memories. They are my "cigarette breaks." While I often run on adrenaline, as if the whole of D.C. were a pool that has been electrocuted, during pumping breaks I slow down. It's work to balance the city's whirring with the soft skin and warm smell of a baby. It's art to enjoy either of these on its own—let alone both together. The lactation room is a room to myself, a moment alone, a chance to pause.

Back at the summit, I clap for the panel I missed. I sit next to the Madonna—she has saved me a seat in a discussion group. Her hair tickles my cheek. I imagine a news story about John Kerry scratching his head the next time he delivers a speech, and the rumors that follow. About how only I will know it started with a hug between my son and me and then the quick leap of bugs from one curly head to another.

Later that night, my husband runs a metal nit comb through my hair. When he doesn't see jumping throngs of lice I make him look again. At work the next day I scratch my head in a bathroom stall and clasp my hands to avoid scratching during meetings. I make my husband do another lice check. We shave off our son's mop of curls. Everyone thinks we are preparing for the hot D.C. summer.

"You don't have lice," my husband tells me yet again. I glare at him, too afraid to let down my guard.

It takes a couple of days for the phantom itching to stop, for me to realize we are lice-free. We always were. I go into the day care main office and ask about the lice.

"Lice?" the director asks. "We put those signs up two months ago. You only noticed them now?"

The Secret Mice of the Smithsonian

Three mice live in the Smithsonian National Air and Space Museum. The mice are cast in silica bronze, surfaces textured to catch the light. Each one is the size of a baby's fist, with a sharp nose at one end and a long tail curved into an S at the other. Perked up ears listen to the tide of tourists that push through the exhibits every day of the year except Christmas. The **first mouse** I found was below a water-cooled engine designed for bombers in World War I and then used to power early mail planes. The mouse is a tiny piece of art, and I want to know its story.

I am on maternity leave. As my return to the office approaches, I venture farther into the city. We moved to D.C. when I was four months pregnant with our second child and unable to do much more than get used to my new office. With the small furnace of my baby strapped to my chest, I study the wild field mouse among the icons of technology that have pushed our boundaries of exploration farther from Earth.

In the liminal space of maternity leave, I mostly cocoon. Quiet days with Cluck fill themselves out with neighborhood walks. I bask in the sweetness. For my first baby, three years ago, I had no leave. I still carry that deferred exhaustion. Thoughts about my future and career linger in the periphery: where we will live, what we will do, if it was a mistake to exit our academic careers. I have just a year for my congressional fellowship, though the clock has paused for these three months.

If I don't sleep when the baby sleeps, I write. When I start a paragraph, I don't know where it will take me. It's a process of discovery. In contrast, writing for Senator Franken is a structural puzzle of distilling information—staying sparse without seeming skimpy. Every word connects to an action: a position to take on a bill, a question to ask in a hearing, a point to make in a speech. The groundwork for confrontation. I had no idea what I was getting into coming to Congress, but I love it, actually.

I know I won't have time for my own writing when I go back. Jane Hirshfield says, "One of the jobs of all art is to look at what the central focus of the culture is not looking at." I want to continue looking for the other world inside our world: the fact of spirituality in our mundane lives. I am talking about wonder. The mice in the museum are clues as to how to do this, to see possibility. To play.

The mice sculptures, I find out, are the signature of the museum's blacksmith and machinist, Chris Modla. Modla weaves period ironwork around artifacts, pitted and finished to capture the light and to take the touch of millions of visitors in the second-most-visited museum in the world, after the Louvre. He designs interactive levers and bags that visitors slam, grab, and punch to experience the principles of flight. He constructs, and repairs, metal and plastic pieces of exhibits: railings, graphic panel supports, fittings, and signs, "made to last almost forever."

A couple of weeks after I return to the office, I meet Modla on a Friday before the museum opens. Moving through D.C. without the baby, and all a baby needs, I am nimble and a bit empty. I admire the planes, spacecraft, and missiles that dangle above. Modla keeps rotating our positions to place himself to my left. Finally, I realize he is deaf in one ear.

Modla is a former naval officer. He tells me about competing at pole vaulting in high school and doing ballet in college. His posture retains the look of a steel rod running through his center. He still dances. "Ballroom is a struggle to constantly push yourself to get better each time." He does international standard waltz, Vienna waltz, foxtrot, quickstep, tango. He tells me that on the dance floor, "we create shapes in space," as if his swirling feet were charting equations on a blueprint. I consider that research isn't something that just happens in a laboratory or at a university.

If Modla felt lost during his career, he has shed the uncertainty from his narrative. After the navy, he completed a master's degree in physics at William and Mary. It was while living in Williamsburg, Virginia, a historic town, that he began blacksmithing. He went on to work at Westinghouse as a computer programmer, ran his own blacksmithing business, taught high school physics, and, finally, came to the museum, where his scientific knowledge and craftsmanship merged. It's as if each experience was exactly what he needed at the time.

Modla's series of disconnected, but interesting, jobs is nothing like the academic dream of one job that lasts forever. It is a career where movement and growth have come through changes in focus. I like the idea that doors will always be open, that my life could include several pursuits.

The **second mouse** is in the "Pioneers of Flight" exhibit, which weaves together the history of war and flight. Modla points to a sign, "Military Aviation": "This picks up the light with all the nooks and crannies." The metal of the sign is textured with glacial striations and dents made by a child-size hammer.

We go into a dimly lit cave marked "Don's Air Service." Orchestra music plays from a black-and-white Mickey Mouse video. We kneel to find the mouse below a display of model airplanes. It is made from the same mold as the first, but this one is accompanied by two pieces of cheese to mark the two interns who helped with the exhibit.

To shape metal, Modla heats it to close to 2,000 degrees Celsius. At these temperatures, the metal becomes plastic and can be manipulated through hammering, bending, and stretching. As steel or iron heat, it turns red. At higher temperatures it turns orange, then yellow, and finally white. Most forging is done when the metal is between yellow and orange.

The craft of metalwork reminds me of fieldwork. I could have found joy working with my hands in laboratory procedures as a postdoc, but I didn't. I found sterilizing glassware and working syringes in a fume hood tedious. I had applied for my postdoctoral fellowship with already collected samples from a colleague. But not having surveyed the plants or dug in the soil, I didn't feel in relationship with the land I studied.

"I try to keep everything consistent, keep it in period," Modla tells me. He points out, again and again, where he has used traditional joinery. Welding joins pieces of metal by melting them together where they meet, in contrast to the more traditional metal joinery of punched holes and rivets, square bolts, or other things analogous to nails. The type of joinery used results in distinct and decorative features.

Modla and I enter the room where Orville and Wilbur Wrights' plane the *Wright Flyer* is displayed. He tells me about forging a replica of the fence surrounding the Wright brothers' home in Dayton, Ohio. A wall-size photograph shows where they lived on Hawthorne Street. The brothers stayed with their father and sister in the house until Wilbur's death in 1912. The fence Modla created is waist height. I touch the top where the black has faded to dull metal from so many clasping hands. I study the visible wear on the wood slats of the benches near the fence, and realize it is from visitors kneeling to gaze toward the plane. No one sits to read the exhibit's wall text.

The story written on the walls asks the question of how midwestern bicycle shop owners were able to design a wood-and-cloth machine that transformed the world. The answer is obvious. Susan Wright, their mother, mentored their mechanical genius. She helped her own father in his carriage shop, and as an adult she designed and built appliances and toys for her household.

The proposed preposterousness of revolution coming from Wilbur and Orville makes me think of an Einstein quote: "In striving to do scientific work, the chance—even for very gifted persons—to achieve something of real value is very small." Every discovery we celebrate represents thousands of false starts, incremental contributions, silent leaps of progress, and brilliance we do not hear about. I am one of thousands of scientists and not one who stands out. But I think Einstein was too narrow in his observation. The judging of a person's gifts is subjective. It is not just the chance of whether a scientist makes a discovery, but also the chance of whether that work is recognized.

Even when work is recognized, credit is not always straightforward. The Smithsonian refused until the 1940s to acknowledge that the Wright brothers accomplished the first manned, powered flight with the *Flyer* at Kitty Hawk, North Carolina, on December 17, 1903. Instead, the museum claimed that Samuel Langley flew the *Langley Aerodrome* in Quantico, Virginia, nine days before the brothers' flight. Langley had been a college professor and secretary of the Smithsonian. In comparison, neither of the Wright brothers had received a high school diploma.

Orville refused to donate their plane to the Smithsonian until the museum recognized the flight at Kitty Hawk as the first manned, powered flight. Instead, the *Flyer* was put on display for the British public at the London Science Museum. During World War II, the museum hid it in an underground vault outside London to protect it from bombing. In the 1940s the Smithsonian relented and signed a contract acknowledging the Wright brothers' status as first in flight. By the time the *Flyer* arrived at the Smithsonian, both brothers had passed away.

The drama did not end with the signed contract. In 2013, the state of Connecticut challenged the Wright brothers' place in history by passing a resolution stating that Gustave Whitehead achieved the first manned, powered flight. Whitehead attempted flight in Bridgeport, Connecticut, more than two years before the Wright brothers' flight at Kitty Hawk. The claim that Whitehead succeeded is based on old newspaper articles, long-lost photographs blown up to more than three thousand times their original size, and the ability of replicas of Whitehead's plane to fly. Proponents of the claim argue that the Smithsonian is again holding tight to a manufactured truth.

I notice a bolt conspicuously missing from an exhibit bench. It could be anywhere in the world, tucked into a box or drawer. "The visitors steal," Modla says. He has learned to make things that fiddling hands can't snap off. Is it vandalism or desire for a talisman? Neil Armstrong took pieces of the *Flyer* to the moon inside his space suit pocket, muslin fabric from the left wing and wood from the left propeller.

Modla tells me, "Studying and learning requires solitude. You have to forgo society to get good, to focus and concentrate. I have endured solitude or else I would have been a jack-of-all-trades and a master of none."

I don't want solitude. I speak too many foreign languages shared by too few: the language of chemical molecules, the language of statistics, the language of botanical fossils. Languages spoken mostly to myself. I don't want to spend more nights alone in a laboratory with screaming machines.

The museum doors below squeak, it is 10:00 a.m., the wave of incoming visitors audible. My BlackBerry vibrates, and my mind flickers to the day's to-do list. The high ceilings and open spaces of the museum carry the loud, excited squeals of children, the patter and stomp of feet, and the voices of adults corralling groups. Like water set loose from a dam, people fill up the halls and exhibits.

Dodging groups of children, Modla stops in front of a model of the USS *Enterprise,* an aircraft carrier. Putting airports on ships was a technological advance of World War I, along with drones and sanitary napkins. Modla's colleague hid a model of a Chevrolet Corvette on the deck of the replica warship. We search but don't find it.

"Is it common to hide things in exhibits?" I ask.

Modla's smile is wide and open, "It is all part of the fun." He winks.

It is too dark to see the **third mouse**. I feel for it with my hand. Above me a mannequin of William Herschel looks through a long telescope. Night sounds click and chirp from a loudspeaker. A booming voice counts stars in German and English. I trace the metal fence's swirls until I find the mouse. My BlackBerry vibrates again. I take a breath and push it out of my mind. I know that once I look at my phone, this world beyond work will fade away.

The mouse is at ankle height. The long tail is curled into an S shape, and the face is a sharp point. I feel the texture that would catch light if there were any. It is a wink to visitors made to last *almost forever.* A reminder to have fun. The world is full of possibility—so am I.

Thank You for Lobbying Our Office

This is what it means to live in the moment: my phone beeped, I saw I had an appointment, and now I am here. Five minutes ago, I was choking back tears—my colleague told me about a meeting with parents who lost their child to a rare disease that has no cure. Before that I was running alongside the senator, who was on his way to a vote, so I could update him on negotiations for another bill.

Business card in hand, notebook under arm, BlackBerry in pocket, I step into the reception area with a smile. I don't know which cluster of people, standing or sitting, hopeful or anxious, wait for me. Two men in suits pop out of their seats, hands held out. Sometimes I take meetings with other staff. Today I fly solo. One teammate is in urgent negotiations with another office. Another is accommodating a constituent who hadn't called ahead.

It is not a game, but it is a challenge to not say what I think, to not react, to maintain calm. To be polite, regardless. Full stop. That doesn't mean I don't have thoughts.

* * *

Thank you for coming.

I can see that you are taller than me. You don't need to stand so close.

Yup, I see that you are older too.

And yes, I am from the state.

Right in here, we have this conference room.

The dog? His name is Blaine.

Yes, I do leave my phone on. I make myself available twenty-four seven to the senator. Other than that, you have my undivided attention.

I was late, but my breasts would have been spraying milk with such force you'd have thought I was setting off fireworks. I have a baby at home, and I needed to pump. So, how can I help you?

You only have twenty minutes left, and I am still unclear on your ask.

Twenty? Because the first ten you spent telling me about the time you saw my boss at a bar fifteen years ago. That is right, I am committed to giving you the whole thirty-minute window.

That is all interesting background, but why are you telling me about this?

I know your name because it's on the card in front of me, but I don't know who you are. I think it would help if you back up and start from the beginning. In thirty seconds, tell me about yourself, your organization, and why you are here.

It's complicated? Well, try me.

Yes. I got your email earlier, and no, I didn't open the attachments. I am taking notes now based on what you say.

Now ask for something—what are you trying to make happen?

That sounds important, but I still don't understand how this connects to our state. Did you talk to the offices that represent the states where this is happening? Yes, they might not be in your political party, but they are human, and this is of direct relevance to their constituents.

If you don't lay out the counterarguments, I will.

I've heard that statistic, but can you breathe life into it? Give me an image or a story that will haunt my dreams until I do everything I can to make something happen. Make the issue stick in my mind like gum on the bottom of a shoe. It's not that I need motivation to work. It is that every moment here is overflowing.

If you go through the Senate's procedures and add up the potential floor time each year, you can see the math doesn't add up. There is not enough time for all the issues to come to the floor. We are drowning in problems that need solutions. Real people need help. Not everyone has the chance to thrive. Too many are just surviving.

I see you are passionate. And, yes, you are right, we could put every drop of our soul into your issue. But we are also concerned about people living without insurance, the rate of veteran suicides, mass extinction of plant and animal species, children who don't have enough to eat, human trafficking, and national security.

It is hard to hear that a long journey starts with a small step. That you never get the full loaf, and you are lucky to get a slice.

There is someone else fighting the change you want, for them what you are lobbying for is scary. Maybe those changes would put them out of business or force them to move out of their family home. I'm not saying you are wrong, but what if they are also not wrong? What if their truth is different from yours?

Every action has trade-offs. That is why government is slow. That is why we have gridlock. Our government was designed to get stuck.

And that is a wrap—I have no idea where I need to go, but my phone will tell me. I will find out what it is when I get there.

Thank you for coming. Now I have your face and your name attached to this issue. Now the numbers tell a story about actual people and communities. Putting numbers into story form: that is ammunition in politics.

Congressional Lactation Diary

APRIL 1, 2013: I am on the verge of tears and half naked in the United States Senate. My pump won't work. I don't know what to do. The lactation room has a chair, a fridge, a sink, beige walls, and a door that locks. I checked in with a nurse at the front desk, but I can't ask for help with this. I hear metal curtain rings slide in the room next door where sick people are lying down.

I don't know if I can do this job and be a mother, never mind keep up milk production. But I can write everything down. This will be my lactation diary.

I just took the pump completely apart, put it back together, and bingo. It is a trick I learned with mass spectrometers in the laboratory. Sometimes, if you can't figure out what went wrong or why—just start fresh.

When I walked into the building this morning, electricity shot up my legs. Each *click click* of my heels sent a shock. The black veins and flecks in the marble walls don't match from one block to another, creating a sense of motion. It is like birds flying in the distance or wind rustling in the leaves. It's like the building is alive.

As a climate scientist I was trying *to make a difference.* But shouting in a room with other scientists wasn't getting anywhere. No one else was listening. I didn't know how to be heard or who to talk to. Here, I am in the middle of it all. BOOM!

A friend from my lab said that my postdoctoral adviser told them I had *gone off the deep end.* She was as likely referring to my having a second child as to my work in Congress. When she congratulated me on my fellowship, she said that if I left, *I could never come back*. But does that mean I'm no longer a scientist?

Ounces: 9

APRIL 2, 2013: To explain my office to Byrd, my two-and-a-half-year-old, I nicknamed my direct boss the Pilot. I told Byrd the Pilot comes to work by helicopter and climbs down a rope from the roof of the Hart Building.

Our office is a loop split into two sections with a stairwell and bathroom in the center. On one side is the communication team, on the other the legislative team. These are the two sides of the senator's octopus brain.

The Pilot and I are the environmental policy tentacle—the biggest issues are in energy and agriculture. We keep up an ongoing dialogue through our shared gray cubicle wall. There are also tentacles for judiciary, health, education, economics, infrastructure, and tribal issues. Each tentacle sifts information and passes on critical bits in a distilled form. What gives me power, and what most terrifies me, is making decisions about what information *not* to give the senator. Giving him everything would render it all meaningless.

To arrive at thoughtful strategies and crisp messages requires research, calls to experts, and follow-up with stakeholders in Minnesota. Then we debate each other in the office. All of this is packaged into memos sent home with the senator in a thick binder each night. In the morning, we pull everything apart on the blue couches of his office for another round of deliberation.

Ounces: 8

APRIL 3, 2013: I know nothing about politics, never mind the specifics of Senate procedure. The only politics my family talked about at the dinner table growing up were hyperlocal: a teachers' strike, the school board and the union, Mayor Buddy Cianci and corruption, the Mafia organizing trash collection, the connection between the mayor and the Mob, and the eventual FBI takedown of the mayor. As a Canadian citizen, my father couldn't vote in

U.S. elections and was ambivalent about becoming an American. Now, he seems slightly disgusted that I would involve myself in this way.

Ounces: 9

APRIL 4, 2013: When I ask the Pilot for career advice, he quotes Joseph Campbell: "We must be willing to let go of the life we planned so as to have the life that is waiting for us."

A coauthor from the research world, from which I have been exiled, just sent edits on a paper. I wrote it after moving to D.C. in the weeks before my fellowship started. It has taken months to get feedback from all the coauthors—a mix of people with varying levels of connection to the work and to me. I no longer have a source of funding to pay for my time—or to hire a babysitter. But my research results are *my responsibility.* If they are going to make it out into the world, it will be up to me. Do I steal time from my family on the weekend? Steal time in the middle of the night?

When I think about walking away from my career in science, I hear the fabric in my rib cage tear. Campbell doesn't touch on real-time angst.

Ounces: 8.5

APRIL 5, 2013: I was the first chance for the senator's staff to show off their lovely parental leave policy. They threw me a baby shower after knowing me for only a few months. No one questioned my ability to be part of the team. Their leave policy didn't apply to my fellowship, but I read it to see what it would feel like if it did. I negotiated with my program to take a maternity leave. They were okay with anything as long as my time off was unpaid.

I bring work to the lactation room. It's quiet here, and that makes it a good place to write speeches for the senator. *Need talking points? Find a lactating mom.*

The Pilot didn't understand what I meant when I told him I have to "pump," but it doesn't matter. He trusts me. He wants me here. How do I know? He gives me more and more demanding work.

When I pump during the workday, or if my BlackBerry goes off when I am home, I feel the tug of a giant party going on without me. My leaving the office at 5:15 most nights is not normal. Other staff meet over frozen yogurt to talk about bills. But I have to get to day care before it closes at 6:00. My hours were the only detail that gave anyone pause. While the office agreed, and they seem fine with it, sometimes I am not.

Ounces: 8

APRIL 8, 2013: I just met with a group of pastors. So many we couldn't fit into a conference room and had to stand in the hall. They were hip: a dove tattoo on a foot, thick oversize glasses, skinny jeans, nose rings, cotton scarves, bruised-leather flats. They wanted to talk about climate change.

Each moment here comes at me, sometimes smacks me in the face. There is no time to look up information before taking a meeting. I often have no idea who the meeting is with before I walk in. There are too many and always more to squeeze in. I have so much to learn, but I stick to the thirty-minute rule. No matter what, watch the clock, wrap it up. I could burst with unasked questions.

This week: a confirmation hearing, federal regulations on fracking, and the Hot Dish Competition. All those fracking chemicals . . . I am at home in the language of organic chemistry.

The senator started the Hot Dish Competition to bring elected Minnesota officials from both parties together. Each official makes a traditional oven bake, many include canned creamed soups, Tater Tots, and/or marshmallows. Electeds show off their regions by cooking with bear or moose or Spam. It is hopeful: those who eat Spam together solve problems together.

Ounces: 9.5; Push-ups: 1 (that is being generous as I collapsed halfway through)

APRIL 9, 2013: I staffed the senator on my own for the first time: total failure. It was the confirmation hearing for Dr. Ernest Moniz to be secretary of energy. The room was crowded to the point of distraction. Journalists flanked the audience. Camera flashes shocked us over and over. Observers without seats stood along the walls and pressed in from the hall.

Senators sat around the dais in the order they began serving on the Committee on Energy and Natural Resources. I crushed in with the other staff on a green velvet bench behind them. My phone vibrated with office text chains counting down the minutes to other hearings the senator had to get to that morning.

Dr. Moniz introduced himself, saying that he and his wife had been married for 39.83 years.

The senator turned to me, "June 11th?"

The movements of the crowd rippled like water. I watched the two transcribers, one typing, the other speaking into a cone. A deer in headlights, I shrugged.

He turned back to the hearing.

Friends in other offices texted, "Looking good." Aware that I was on C-SPAN, I had the idiot urge to smile.

Just before his turn, the senator turned to me. "So, is it June 11th?"

My face filled with heat. It hadn't occurred to me to do the math. He waited. It hadn't occurred to him that I wouldn't do the math. Senator Manchin's staffer, sitting next to me, leaned in, "I got June 10th."

"Dr. Moniz, thank you for being with us," the senator said in a serious voice. "You say in your testimony that you've been married 39.83 years. May I remind you that you are under oath?"

Dr. Moniz laughed, nervous.

"So is your anniversary June 10th?"

It was the only moment Dr. Moniz stumbled in the hearing. "It's June 9th. That's in the rounding error."

The joke failed. But it did its job: the senator had the room's attention for his formal line of questioning.

The hearing was long, and by the time I got here to pump, every cell in my body was ready to explode.

APRIL 10, 2013: I just sat across from Gina McCarthy in the senator's office. She is President Obama's nomination to be the administrator of the Environmental Protection Agency. She pronounced "tar sands" as "taa sands," the way kids I grew up with in Providence dropped the "r" at the ends of words ("caa companies," "caa-bin pollution") and added it to words ending in vowels ("pizzer").

The Pilot keeps checking on me: "Do you miss the baby?" "Are you okay being back?" With an iceberg, what you see on the surface is only the top third—the rest is underwater. I am like that, an outer third of confidence. Underneath it's more complicated. But the only thing that has made me want to lie on the floor and sob my eyes out was hearing Gina McCarthy's accent. I had to pull my suit jacket closed to hide the wet marks, my breasts weeping even as I held my face blank.

Before ending the meeting, the senator asked me and the Pilot, "Do you want to bring anything else up?" And just like that I was no longer present as a resource, but fully drawn into the circle with a badass woman likely to be in Obama's cabinet.

Ounces: 9; Push-ups: 1

APRIL II, 2013: I'm working on a speech for the senator, on climate. It needs to be good, to be amazing. But when my fingers hit the keyboard, the words evaporate.

The rules of writing a memo are brevity, political context, and a clear ask. But memos are internal documents and speeches are external. I don't know how to write a speech. I certainly don't know how to write a speech for someone famous in showbiz.

Ounces: 8; Push-ups: 2

LATER: I've gone through the looking glass! I watched videos of the senator until I felt a quickening. I ran out with a pen and a notepad and sat in the window behind a large potted plant a few floors up. At first a stream of consciousness came out, but then the words started to flow, and the words flowed *in his voice*. It was like learning Italian, and how speaking a different language comes with a slightly different personality. The process, the bodily function of writing the climate speech, was the same as writing fiction. A letting go and an opening to a state of flow.

Ounces: 7; Push-ups: 2

APRIL 15, 2013: The building has been on lockdown since bombs went off at the Boston Marathon. It is horrible. And yet, I couldn't stop focusing on my body being trapped without access to pump. I ran here as soon as I was allowed to leave our office. I wonder what made the sergeant at arms think we are in danger. I worry that my milk will taste of fear when my baby drinks it tomorrow.

Our safety precautions make me jumpy. We have a seventeen-page spiral bound *Emergency Action Plan* with twenty-one glossary words and thirteen acronyms written specifically for our office.

Each morning I walk through a metal detector to enter the building. "Don't slow down," the officer says as if the machine is a dog that sniffs fear. The presence of armed men and the act of putting my bag through a machine make me feel like I have done something wrong. An alternative perspective is that I have something to fear from others coming into the building. Great options.

In my bag I carry lunch, the makeup I apply for the final transformation from mother to staffer, and fresh baggies for the milk I pump. *Errhhh urg, errhhh, urg.* God, the noise of this pump drives me crazy.

Ounces: 9; Push-ups: 3

APRIL 17, 2013: To protect our office's privacy, I enter Senator Franken's number into my BlackBerry as "Skipper." I didn't coin that nickname. I copied the Pilot. Our phones sit out during meetings so we are available—fires spark all the time. The inner workings of our office, even texts and casual conversations, are made intimate by the constant threat that they could be taken out of context and used to attack us.

Ounces: 8; Push-ups: 3

APRIL 18, 2013: The Pilot and I got coffee like we do every day. The walk to the cafeteria is our check-in. We talk through memos, strategize upcoming hearings, and argue. I like the arguing. It feels good to not agree, for someone to challenge me. It makes us both feel smart. And, most important, it prepares us for the senator, who will ask about a detail, or for more background, or dig into underlying assumptions, and expect a comprehensive report. The senator gets hangry for information—he eats whatever food the scheduler puts in front of him, but when he asks if "hole" is the technical term in a fracking operation, my mind goes blank, and his exasperation cancels out all my research for the memo. One of the first things the Pilot taught me was to say "I don't know" if I am not completely sure, which seemed obvious, but now I get why one might reach for an answer just to get out of the hot seat.

The Pilot and I trotted down marble steps more worn in the middle than on the edges. The staircase is a battle map of senators running to vote, aides racing to committee hearings, and lobbyists stomping out their frustration. A reminder of how many have come before us, how many will come after.

"This gridlock," I hissed. We are stuck inside a cliche. I came here to do things, and it seems impossible that we'll accomplish anything.

"You have to understand. This is a civil war," the Pilot said. I must have looked alarmed because then he added, "A *civil* civil war."

"We have rules and procedures instead of guns and cannons," he said. "We are fighting for completely different visions of the world."

"A war?"

"A war that is not meant to be won."

I nodded, feeling, as I often do, like the younger sister watching the cool older brother.

Ounces: 8; Push-ups: 3

APRIL 22, 2013: My name is in the official *Congressional Record*!

MR. SENATOR. Mr. President, I ask unanimous consent that Anna Henderson, a fellow on my staff, have privileges on the floor during today's session.

THE PRESIDING OFFICER. Without objection, it is so ordered.

The room was regal, with gold wallpaper and a blue-and-red carpet. The presiding officer punctuated the passing of time with a gavel. Consciously making history is not about words. It is theater. I watched the senator, as the ghost of Christmas yet to come, tell Scrooge about climate change.

High school–age pages in black suits slumped around the presiding officer as if waiting at a bus stop. A scribe madly hit keys on a typewriter worn around the neck. Only a few of the fifty desks in the room were occupied. Senators are too busy to listen to each other. They come to vote or to give their own speeches. While the room was mostly empty, the cameras were on, and Senator Franken spoke like he had an audience.

This is what I came for, to be part of the push for climate action. But then, just like that, the Senate moved on to discuss American companies moving overseas.

Ounces: 9; Push-ups: 3

APRIL 25, 2013: An oil and gas lobbyist was going on and on about how proposed regulations would ruin the industry. I questioned her claim that the companies she worked for couldn't disclose the chemicals used for fracking. "It's really complicated," she responded, and handed me a glossy one-pager.

"I am up for hearing about it," I said.

"Um, I don't really know the details," she said. The idea that *information is power* is not just about trading rumors. People seem surprised to find a scientist on the Hill. But science is my superpower. I have the fundamentals to learn whatever topic erupts.

Ounces: 8; Push-ups: 3

MAY 1, 2013: I was told that thinking of my baby would help my body let milk down. It's not true. I don't look at his photo or the thirty other smiling babies pinned up on the bulletin board. I think of nothing at all. This works because my secret desire, in my heart of hearts, is to be alone in a silent, completely blank room. Not so far from the lactation room.

Once body mechanics take over, I look around, write in this journal, send emails, read magazines, or page through *The Motherly Art of Breastfeeding*. Yes, this is a motherly fucking art.

Ounces: 8; Push-ups: 5

MAY 6, 2013: In my cubicle I hear paper rubbing on paper, the gurgle of water flowing into a mason jar, a colleague on speakerphone, the click of a spoon in Tupperware. I hear Blaine, the office dog, pushing a plush hedgehog around with his nose. He sneaks up on me when I take out my lunch.

I can identify the sounds of my colleagues the way I knew the sounds of my six siblings growing up. Each has their own soundscape: the way they run, the swoosh of a suit jacket, a loud exhale, the smack of a folder on a desk. I wonder what iconic sound my colleagues know me by. It occurs to me that it might be a smell and not a sound—I carry the perfume of slightly soured milk.

Ounces: 9; Push-ups: 5

MAY 14, 2013: I have the same conversation over and over:
"What do you do?"
"I work in the Senate."
"For who?"
"Senator Franken"
"He must be funny all the time."
"No."
"No?"
"No, he is never funny."
"Al Franken? He's not funny?"
"No. I've never met anyone more serious," I say.
After the Moniz hearing, the senator's chief of staff came

running. Without raising her voice, she can be absolutely terrifying. Her disappointment enough of a blade, she demanded to know why we let Skipper tell a joke. It was our job to stop him.

When I started in the office, I was told to manage the senator. I balked. "How?" I asked a young staffer who fell in step with me at the Metro. "Use your stink eye, moms are experts at that," she said. I don't even have a beta version. Babies and toddlers don't really take direction. I am more the fall-to-my-knees-ask-what-is-wrong mom. The senator does talk fondly about the time a staffer passed him a note telling him to *stop being an asshole.*

Ounces: 8; Push-ups: 5

MAY 21, 2013: Literally all I did was drop some paper in a box. But the whole subway ride from the Senate to the Capitol I had to force breath down into my bones. I took the lead on a bill for the first time. And it's a big bill—a billion dollars—it will likely pass as a title in the farm bill.

The farm bill sets the rules on which farms, supply chains, and corporations are built. It is a central point of control for environmental policy. Working on it is like being caught in a giant wave. It's much too big and out of control for anyone swept up in it to do more than just stay afloat. I have to remember I'm coming in at the middle of the story. This is not the theoretical world of good ideas. It is the concrete world we have to work with.

We meet with any farmer or environmental advocate or industry lobbyist who wants to talk. They tell us what they need and want, and how to improve the bill. But trees don't have legal standing in politics—rivers don't have personhood. To have a voice here requires one of two things: economic value or a good story.

Ounces: 9; Push-ups: 5

MAY 24, 2013: This is what a climate discussion looks like in the Senate: about thirty staffers, maybe two from each office, about fifteen of the fifty Senate offices represented. We sit at a tall table with low chairs, normally the press's feeding trough during hearings. I feel like a child reaching up and over the table to pass out my business cards. It's recess and the attire ranges from jeans to cocktail dresses. Some people wear suit jackets with jeans, a sort of D.C. mullet with business on top and casual on the bottom.

We talk about emission trading programs. China has a pilot. Europe wants to take the lead. Australia passed legislation. Policymaking, like the scientific method, includes observation, testing, and lots of skepticism. The conversation is exciting but also abstract because the value of reducing pollution—or suffering its consequences—is both concrete and existential. This is the type of conversation I wanted to be a part of in academics but didn't know how.

Ounces: 7.5; Push-ups: 7

JUNE 10, 2013: What I love about this place is that everything ends! For better or worse, the farm bill passed in the Senate. We had two hundred and something proposed amendments. We raced to decipher what all of them meant and make recommendations for how the senator should vote on each one. At any moment a deal could be struck and then a package of amendments would come up for a vote. Amendments are not written to be understandable. Most add to existing laws, building on them like barnacles growing on rocks. If a new law guts an old law, it's more like a hermit crab taking over an old shell, filling it not only with new life but also with a new species.

It doesn't make sense to me that laws can be made or changed like this—there's no way the elected officials understand everything they are voting for. But then, neither does the public. The only accountability is if the media cover a law's passage, or if NGOs grade the elected officials on their votes.

In my life as a scientist, a day of research in the field became years of lab work, data analysis, and peer review. The research process went on until what was once passion made me want to puke. While the movement of Congress is painfully slow, when something happens, it is rapid-fire.

Ounces: 9; Push-ups: 7

JUNE 12, 2013: I gave money to a man singing at the top of the escalator in Dupont Circle. It was like he was building a cathedral over us with his deep voice. I was late. I couldn't stop, but I stuffed some bills into his jar and rushed on. What would it be like to linger?

Work, then home, then sleep and nursing. Very little downtime. This room, this notebook, my bit of earth.

Ounces: 8; Push-ups: 8

JUNE 20, 2013: The House voted on the farm bill, and it failed. That fight isn't over.

Ounces: 7; Push-ups: 8

JULY 1, 2013: I worked on my science colleague's edits and submitted our paper to *Geochimica et Cosmochimica Acta*. I feel numb. Finishing the paper was a kind of makeup sex—or maybe more like pity sex.

On Saturday, while Dan took the kids to the zoo, I worked at the dining room table. A flood of emotion overtook me. At first, I was excited. Then angry. Then a cold draft of fear worked its way into my gut. I still feel it—the fear. I don't know who I am outside academics. I am utterly lost.

Ounces: 9; Push-ups: 10

JULY 9, 2013: We had a real fight. Other staff turned around to watch. "I thought we fought for climate, not coal!" I growled at the Pilot.

When I came around, it wasn't from his counterpoints. It was from meeting with twelve rural mayors oscillating between desperation and ferociousness. We sat around a table, and they told me how their Main Streets depend on cheap electricity. After that meeting, I drafted a letter for the senator asking the Environmental Protection Agency to reconsider regional haze restrictions on North Dakota coal plants that serve Minnesota. Haze regulations protect the vistas of national parks and wilderness areas from smog.

The same chemical compounds from burning coal contribute to haze, climate change, asthma, lung cancer, strokes, and congestive heart failure. When the narrative is about health and climate, we fight to tighten regulations, and, at the same time, we fight to loosen regulations when the narrative feels disposable. My letter basically boiled down to the argument that a clear view of the world isn't all that important. The approach is about being realistic: a mix of compartmentalization and hypocrisy.

I feel like a bad environmentalist. Dealing with climate change is not a straightforward equation. Meeting with the mayors brought humanity into a conversation more often framed around technology. Today, I didn't know what it meant to be brave.

Ounces: 9; Push-ups: 12

JULY 12, 2013: The Senate green room is a bipartisan space behind the hearing room. The room is not green. The walls are lined with brown legal books that look like encyclopedias. Today was the first time I've seen anyone use one of the books. I found a twenty-dollar bill on the floor and showed it to the Pilot. "Keep it," he said.

I must have made a face because he grabbed a book off the shelf, opened it, and laid the bill between the pages. The door to the room creaked open, and we exploded with giggles as he stuffed the book back onto the shelf.

How long until someone will come across the twenty? Laughing with someone makes them feel like kin.

Ounces: 7.5; Push-ups: 10

JULY 17, 2013: At the staff meeting, the heat started in my feet and worked its way up until I was wiping sweat from my forehead. Senator Stabenow's office had asked us to work together on a bill supporting industries that make things like plastic from corn. I needed to ask Senator Franken if he wanted to sign on, but a colleague was taking forever going through her week's accomplishments. A cat laying dead mice at his feet. I'd memorized my concise pitch and wanted to get it over with.

The scheduler burst in, out of breath. "He's on the phone!" she said.

The senator let out a sigh of relief, or resignation, and walked out.

Speaking of sweat, I am drenched. Every time I pump (or nurse) this happens, as if milk involuntarily spraying out at random were not enough. Thank goodness for the sink in this room.

Ounces: 7.5; Push-ups: Shit, have to run

LATER: The lead office on the bill called, insisting we make up our mind. I asked the scheduler for time with the senator. "Come at two and you can have ninety seconds," she said.

At ten past two, the head of communications walked out of the senator's office holding up a caricature drawing. "It's for a silent auction," he said. I've seen Skipper draw other senators in hearings and offer them the picture afterward. The way the dais is arranged, his gaze rests on members from the other political party. They always decline.

Usually, there is a crowd on the couches in the senator's office. But today he was alone at his desk. I knew there were three possible outcomes:

a. He says, "Yes, great idea!" Then we discuss babies or dogs for thirty seconds.
b. He digs and digs, asking questions until I say, "I don't know. I'll find out for you."
c. He says, "No," and I know I have done something wrong.

I had vetted the bill with the head of policy, the head of the state office, the head of communications. It was a no-brainer, a win-win. Besides, I had rehearsed the pitch in a bathroom stall and mumbling under my breath in my cubicle. *I had this.*

He said, "No."

I smiled and said, "Thank you." I meant it. He is the boss. His own person. We have a leader. I walked out feeling good because I knew that he is responsible for our work. I did my job, but he makes his own decisions.

Ounces: 7.5; Push-ups: 15

JULY 18, 2013: Tina Fey said that the rudest thing you can ask a woman is, *How do you juggle it all?* What I hear—from my hairdresser, a lobbyist, a stranger in the grocery store, my parents, is that *it's hard to work and have children.* There is also, *It goes so fast, appreciate every second.*

I am supposed to appreciate running to pump because a large group is waiting in our office. Or maybe they are referring to how I'm self-conscious that I still look pregnant. Or that yesterday, all in one day, I wrote a speech, staffed a hearing, and had two children vomit in the bathtub. I especially appreciate cleaning up vomit.

Their comments are autobiography masked as advice. Nostalgia for memories they have compressed into Disney highlights.

It wasn't part of being a nice girl to get angry, but it sure as hell is part of being a mom.

The problem is that work is the break from home, and home is the break from work. I need more recovery than that. I want creative time for myself. And I don't want any fucking advice.

Ounces: 7.5; Push-ups: 12

AUGUST 26, 2013: *Geochimica et Cosmochimica Acta* rejected my research paper and gave instructions for resubmission in the same paragraph. I put so much of myself into those data. "Fifty Ways to Leave Your Lover" doesn't include what to do when the lover is science data.

Paul Simon, write a song for the lactating moms. Please?

Ounces: 9; Push-ups: 15

AUGUST 28, 2013: One moment from my training replays in my mind. Judy Schneider from the Congressional Research Service spoke to our group of AAAS fellows, spit flew from her mouth, and all of us leaned closer. We wanted to absorb everything she had to give.

"We all know certain things should happen, both sides agree on them, but they don't get done. Or one party is in control and they don't get done." She was almost shouting when she asked us,

"WHY NOT?" I was thinking of Obama's failed climate legislation. She didn't wait for an answer: "I'm gonna tell you why."

We held our pens at the ready.

"What happens in Congress is the result of three things: Policy! Politics! And People!"

I had thought that science was pure—above mean girls and value branding. But research is made up of people and their relationships, just like everything else.

I am making a decision. I have believed that I owe science, and the world, whatever it takes to publish my research. I believed the rules of academics were the rules for my life. Sometimes you can't figure out what went wrong or why—just start fresh. Riptides are full of postdocs, adjunct professors, and visiting scholars. I'm going to swim parallel to the shore and out of that world. *Policy, politics, and people.*

Ounces: 7; Push-ups: 20

SEPTEMBER 9, 2013: The Farmers Union came in with a group so large we had to reserve a special room—the fate of the farm bill is still uncertain in the House, so the churn continues. The senator started with a joke. "I grew up in St. Louis Park, Minnesota. You know where I thought food came from?"

They called out, "The grocery store." This wasn't the first time he'd told this joke, and anticipating the punch line created a sense of cohesion in the room.

"I'm not an idiot," Franken said, and the group roared with laughter. "I knew food came from farms." The joke acknowledged the obvious. He didn't grow up on a farm, and yet here he is making farm policy. Emily Dickinson says, "Tell all the truth but tell it slant." His joke was a slant way of saying, I'm an idiot with a tractor, but I'm not an idiot overall. It broke the tension, released nerves, and got the meeting off to a good start. I saw the utility, but it didn't make me laugh.

Throughout each day, the senator enters situations, and like a Lydia Davis short story, he takes reality a step further to highlight the absurdity in the mundane. Making jokes is a kind of tic, an ongoing crossword puzzle, a way to navigate the world.

I've come to see his jokes as a three-legged stool: delivery, surprise, and contrast. A three-legged stool used to control the mood of a room, to change the subject, to focus an audience. With his staff, the jokes are wordplay or silliness. But facing outward, they are calculated for impact.

SEPTEMBER II, 2013: The senator's grandson came in this morning, a tiny bundle of baby smells. The body man, the Pilot, the press secretary, and I waited while he gave the baby a bottle. Everyone but the senator was antsy. We were late.

On the internal train to the Capitol, we talked about what babies need.

"Touch," I said.

"To feel secure," the senator said.

The Pilot, a forty-year-old bachelor, said nothing.

"That was a nice way to start my day," the senator said. He sat, content, smiling.

It is. Since that moment, I have been thinking about how when I get home, I will throw my suit and BlackBerry on the bed and fall into the cuddle puddle of trains, pacifiers, snot, and little boys. I will leave the ringer of my BlackBerry on, but it will still be great.

Ounces: 7.5; Push-ups: 20

SEPTEMBER I2, 2013: The House hearing room has high darkwood ceilings and bold green walls. We camped out at a plastic folding table with a chipped coffeepot and a stack of paper cups. Climate experts sat at one end of the table and congressmen at the other. The only women were staff. The press waited in the hall. Outside of any need to perform, the discussion meandered, open-ended, the lens more on making sense of impending doom than on strategy.

The men in portraits on the wall looked much statelier than those present. Everything always looks more imposing in retrospect, set in a frame. In the moment, it is not clear if the flesh-and-blood men in the room are making history or just spinning their wheels. Maybe the problem is that they are all men.

I slept through the night for the first time! I'm going to try a second set of push-ups today. BlackBerry vibrating . . .

Ounces: 7; Push-ups: 30

LATER: We have a bill on the Senate floor! As chair of the Energy Subcommittee, the senator has led the amendment process and hearings. Something is going to actually happen! The senator delivered a speech that I helped write promoting energy efficiency. It is bipartisan because it saves money. A win-win. But HVAC systems, factory processes, and insulation are not sexy. No one gets excited hearing "the Energy Savings and Industrial Competitiveness Act." The name doesn't roll off the tongue.

Ounces: 7.5; Push-ups: 10

SEPTEMBER 16, 2013: We don't have much information about what happened at the Navy Yard, but they are telling us to freak out. A shooter on the loose, or maybe several shooters. News headlines, gossip, and confusion trickle in. People are hurt, no names or faces yet. I ran here when they broadened the lockdown from our office to the entire building.

My milk isn't coming. I need to trick my body into feeling safe. It is the same all over: Aurora, Sandy Hook, the Boston Marathon, Santa Monica. But right now I need my milk to let down so my baby has food for tomorrow.

When I ask the Pilot if he is afraid, he reminds me that he grew up in a revolution. His family came to the United States when he was in middle school. In quiet moments he gets a faraway look and tells me about his parents pushing furniture against the door as a barricade. Having survived very real violence, he does not react to every threat the way I do. This grounding is part of his physical presence. A grace in the way he moves through congressional chaos.

Ounces: 3; Push-ups: 30

SEPTEMBER 19, 2013: The climate and energy bill has been over-loaded with 125 amendments that will sink it like stones in pockets. It's what people here call a Christmas tree bill. All summer we've carefully negotiated bipartisan hearings for proposed amendments. Anything controversial was thrown out. The open process on the floor allows for proposals equivalent to fighting words: *Keystone Pipeline, carbon regulations,* and, of all things, *Obamacare.* This place is all piss and vinegar. I hear from older lobbyists and staffers about how things used to be with earmarks. If we could offer senators money for a teapot museum, or whatever pet project would help win votes in a tough district, they would get on board. Pork could make a bill personal—it could harness the will needed to pass legislation.

The Pilot is sure we won't get another chance. That was it. We could have had action on climate. Instead, we will freeze the government. The media aren't telling that story.

Ounces: 3; Push-ups: 30

SEPTEMBER 24, 2013: It was one of many meetings in a breathless day, but the call changed everything. I talked with city of Minneapolis staff who had worked to pass an ordinance requiring building managers to track and share data on their energy use. If people don't have information about how much energy individual buildings use, owners won't make improvements and renters won't have the information to choose a building with lower utility costs. Some senators bring up the power of "market forces" as if they are the hand of God. The logic goes that if there are real savings to be had, people would know and that knowledge would influence their behavior, correcting problems in the system. Senator Franken repeats, "We cannot manage what we don't measure," like an incantation. If we can make the forces at play visible, we can change the market.

I asked the city staff about the process. *How did you get something done?* They invited stakeholders to meet. Everyone came: those in support and those opposed. They sat together. They talked. People spoke, compromised, and designed an ordinance. What a fairy tale.

People in D.C. don't show up for policies they oppose. Working in gridlock is like being a bird flying into a glass window over and over. In the definition often credited to Albert Einstein, insanity is doing the same thing over and over and expecting different results. Each time we come at the window, we find a slightly different approach, but that is exhausting to keep up.

I want to be in a community where people show up. I want to *get things done.*

Ounces: 4; Push-ups: 25

SEPTEMBER 30, 2013: Only essential staff will be here during the government shutdown. No one will be paid. Al and his wife, Frannie, offered loans to staff for their rent. The line between work and personal is falling away.

Ounces: 3; Push-ups: 35

OCTOBER 3, 2013: The first warning came from a voice on the intercom: "SHELTER IN PLACE, SHELTER IN PLACE." I was in our large conference room with a group of environmental advocates. We had gotten as far as shaking hands, passing around business cards, and them starting in on coal ash. I was distracted because I had forgotten my phone at my desk.

BEEP, BEEP, the intercom blared. The doors to the conference room opened and other staff came in. I jumped up to run for my phone. "Everyone in!" the chief of staff shouted from the door, and the tide pulled me back to my seat.

The smell of my boys was still fresh on me from the morning drop-off, when my three-year-old clung to my leg, sobbing, "Mama, mama." If this was my last hour of life, my last few minutes, I wanted to tell my husband and boys I loved them. I turned to borrow a colleague's phone, but I realized that if I spoke, the dam would burst. The milk in my breasts stung.

"CLOSE, LOCK, AND STAY AWAY FROM EXTERNAL DOORS," the voice on the intercom blared. The seats around the conference table were full. The group of LGBTQ activists who had been meeting with the senator stood against the wall. Several of our staff crowded in around them. The new AAAS fellows had eyes the size of silver-dollar pancakes. One bit her fist. Another's face contorted into dry sobs. The House had been in session, and we heard the intercom tell everyone in the House chamber to hit the deck. I pictured the representatives lying on the floor in their suits.

"SHOTS FIRED. SHOTS FIRED," the intercom told us.

"I heard gunshots before I came in," a colleague told the room.

I thought of tipping the table and hiding behind it in case bullets came through the wall. The room felt too hot. I was having trouble breathing. The air was charged and sharp. We were sardines crammed too tight in a can.

Congress is a tinderbox. I've worked in the Senate for only a year, and we have had three lockdowns. Plus, the bomb threat last fall that turned a half-hour interview with the Finance Committee into an hour-and-a-half hangout.

I looked up at the senator. He had his palms on the table. His face was calm. I could see his mouth moving. He wasn't cracking jokes. He was asking questions. I focused on this. I felt the table under my hands, the chair against my body. I looked up at the pictures of past Minnesota senators on the wall, heard the static of the alarm system. I watched the hand-biter respond to the senator's questions. She smiled, arms resting peacefully in her lap.

One by one, the senator drew us into the present, into our living bodies crowded together in a safe space. I realized that more than pinching my leg or clenching my teeth, listening helped me stay present. Whatever might happen, in one minute or one hour, for this moment, I was here.

When we were allowed out of the office, I saw the smashed car out the hallway window. A mother, a Black woman, Miriam Carey, had been shot numerous times by the police. Her baby was in the car. The police perceived her as out of control. The only shots fired had come from the police. The House members gave the police a standing ovation for keeping them safe. But the only threat was maybe in the story told.

Officers carried the baby into the building, here, to the nurses' office. *Her baby was in the car.* I wonder what happens to her now, a baby without a mom. Each time I think of her, a needle in my heart.

Ounces: 3; Push-ups: 35

OCTOBER 23, 2013: I dreamt I had a book made of stones, and I wrote on it with water. Nothing lasts. I'm at the beginning stage of cynicism: I love my job and I believe in our work. I don't want to wait for that sentiment to rot. I see how passion curdles with disappointment when colleagues are in Congress too long.

Last night I put in a couple of applications for jobs in Minnesota—where we went to graduate school. I want my children to experience ice skating and kite festivals on frozen lakes. Who knows what will be left of winter by the time they grow up. I miss our community of friends. I'd like to live in a house with a yard.

Ounces: 3; Push-ups: 31

OCTOBER 29, 2013: As far as fashion goes, in the lactation room all that matters is how easy clothes are to take off and put on. A dress without zippers is best. The group that got out of the elevator when I came here could have been headed to prom: ruffles, lace, chiffon, heels, manicured nails, body glitter. It didn't help me feel less like I'm shouting wearing this new dress: leopard print with sequins on the collar.

As a science researcher I wore clothes that could dry if I waded in a river. I shopped at secondhand stores to make do living on a stipend. I got a bezel-set engagement ring so that mud and rocks wouldn't knock out the stone.

Every time I remember I'm wearing bright-pink lipstick, I put a hand over my mouth. I feel like I am cross-dressing. Even though I am a woman, I've never presented as a woman like this. I know I don't feel comfortable, but not feeling comfortable is somehow on the spectrum of normal here.

I told a colleague, "I've started thinking of professional attire in the Senate as anything you would wear to a wedding." She shook her head, "I think it is anything you would wear to a funeral."

I spoke on a panel earlier. After my turn, I wanted to keep talking. It was like the party ended and the music stopped, but I was still dancing. I felt slightly out of control, picked up by the city's frenetic momentum, like I had slipped on Hans Christian Andersen's red shoes. D.C. has gotten into my bones. It's not just that I dress differently—new parts of myself have been set loose. I am not sure I can sustain this.

Ounces: 3; Push-ups: 40 (Two sets of 20!!! No going back.)

NOVEMBER 6, 2013: My friend Katie is in labor in Minnesota—I will hold my breath until it is over. She says they will treat their first kid like it is their sixth. But she doesn't yet know what it will feel like. For me, it was like staring at an eclipse straight on—little black dots bouncing in my vision ever since.

Ounces: 3; Push-ups: 40

NOVEMBER 14, 2013: The stack of business cards on my desk holds the stratigraphic layers of my work in the Senate. Just like geologic samples, with the oldest on the bottom and the youngest on top. The Lamb and Wool Association, the Truth about Snowmobiles, the Real Estate Round Table, Monsanto, the Farm Bureau, the National Resource Defense Council, the Environmental Defense Fund, utility lobbyists, university lobbyists, and on and on.

Issues flare up like geysers and then go silent: agriculture trade groups with the farm bill, propane and natural gas interests for the brief moment before our transportation amendment is cut, perfume makers and chemical industry representatives after the senator mentions green chemistry at a hearing.

Ounces: 3; Push-ups: 45

DECEMBER 18, 2013: I have a video interview for a job as a planner in Minnesota state government! The manager and I had an informational interview last week. We talked for only a minute before he said, "I want to schedule a formal interview." Minnesota feels more like home every day I work in the Senate. I want to go home.

Last day of pumping, last day before the holidays. The baby got his first tooth and bit me—I wanted to nurse for a year, and I did it. I am ready to move on.

Ounces: 3; Push-ups: 60

JANUARY 24, 2014: I got the job! We are moving back to Minnesota. I am so excited to raise my boys with city lakes and the Mississippi River. I have missed the bakeries and the art scene. Rereading the first few pages of this diary made my eyes water. I still feel lost. The neon pink "Fortune Teller" sign on Lake Street flashes in my mind. I am going to have my palm read when I get back.

I was afraid to leave academics. My unpublished research results still choke me if I think about them. But coming to Congress was like giant doors swinging open to let the sunlight in. That is the true measure of a good breakup, when despite the heartache, the world gets bigger and brighter.

I get it now—I am a scientist wherever I go.

Push-ups: 60. Work it like a mother.

IV. Water Czar

At the national level, climate change is an existential problem. At the local level, climate action narrows to the concrete and ordinary: a drainage system in a cornfield, a process at a wastewater treatment facility, a piece of equipment in the basement. All this makes my heart beat faster. I hum learning about local ordinances. I am all ears when farmers gather at a small-town bar to learn about energy efficiency. Returning to Minnesota, I can raise my children on the lakes, ice, snow, and rivers. And my work can make a tangible difference.

Water Is the Medium

Water is the medium by which we experience climate change: catastrophic floods, deathly dewpoints, drought, melting ice sheets, and extreme rainfall. When I leave the U.S. Senate to develop a state climate plan for Minnesota, I don't know that my role will evolve to focus on water policy. That water will become a medium through which I examine power.

My friends in D.C. warn against my career move. Taking a pay cut for a job that doesn't create a path to leadership is risky—foolish even. Coming in at a lower level could limit my potential growth. Technically, three men above me lead the climate plan. This means they represent it at podiums and onstage. As the project manager, I work with about eighty staff from ten state agencies to design policies, analyze data, host public engagement events, and craft the narrative. I moderate debates among agencies, vicious or stubborn or tearful. I hone key messages and turn them into sketches, going back and forth with the graphic designer. My role is beautifully intellectual—I just might not get credit for the work. To protect ourselves and nervous agency staff from legislative attacks, we don't list our names on our reports.

Minnesota passed greenhouse gas emission reduction goals in 2007. The first benchmark is in 2015, a year away. We aren't on track to meet it. Not even close. Carbon dioxide in the atmosphere is 399 parts per million and rising. This is higher than atmospheric concentrations over the past 800,000 years. The ultimate goal is to reduce emissions 80 percent by 2050. That means remaking our economy. There are a lot of ways to do that. The goal of the state climate plan is to come up with the best path

forward—equitable distribution of the benefits without dispro-
portionate distribution of the costs and disruptions.

To me, public service forms the ultimate frontier of entrepre-
neurship. The legislature passes bills in a fever. Each bill holds
a mix of ideals, compromise, and unclear sentiment. Legislators
hand us the blank canvas. Implementing new laws, public servants figure out how to do something that has not been done
before. We make the road map. Our work has the potential to
change everything.

My office was created in the seventies to coordinate the
state's environmental work, and then it was defunded in the mid-
aughts. A skeleton staff kept up the bare minimum of legal functions until Governor Mark Dayton brought back a tentative version of the board in 2013. The skeleton staff left—one fired and
one taking a job elsewhere. We have no institutional memory. My
boss uses a temporary hiring process to populate the office. The
boss is new. I am new. We all are new. The boss promises to make
us permanent within the year.

Our offices are housed within the Whale, the Pollution Control
Agency building. Despite our vestiges of legal oversight over that
agency, we're swallowed by the Whale, which tries to digest us
and spit us out all at once. The Whale's human resources department tells me my job can never be permanent because of how I
was hired. I don't know who to believe. Staff from other agencies
pull me aside to say that they don't want to work with someone
who will not be around long. Many of them have worked for the
state for decades.

The Whale's belly has six floors of gray wall-to-wall carpeting.
Each floor holds its own labyrinth of gray cubicles navigable by
the hot breath of printers and the tangled cords of videoconfer-
ence paraphernalia. Staff set up their cubicles with flowerpots,
posters, baskets of chocolates, teapots, framed photos, stuffed
animals, radios, maps, and stacked folders. Each cubicle is a home
away from home. Office policy does not provide much flexibility:
work happens at work. I spend more time with the people in the
cubicles around me than with my family. My cube neighbors are

a mix of my office staff and staff of the Whale. Constant interaction wears me down. Once, I reserve a conference room to sit by myself.

Governor Dayton keeps announcing that he will end the burning of coal in Minnesota before he dies, *and that isn't too far off.* A cold sweat stands out on his face when he holds press conferences. He has a perpetually waxy sheen. I want to make sure we have a plan for him.

In both academics and the U.S. Senate, I'd seen the world in brushstrokes of good and bad. Growing trees = Good. Burning fossil fuels = Bad. But from my cubicle in Minnesota state government, the problem is more complicated.

Taking stock of greenhouse gas emissions indexes our lives: fertilizer to grow crops, fuel to keep houses warm, combustion engines to drive school buses, concrete to pave roads, energy to operate water treatment plants, refrigerated trucks to transport food. A map of how we live, work, and play. We need to reduce emissions without shrinking jobs or dreams or families. In public meetings, someone inevitably brings up population reduction as the solution to climate change. *And who would control this?* I want to ask.

Most staff in the Whale work on regulations. I learn about our relationship to the forests and the lake-dotted landscape through regulatory permits. Permits facilitate the legal emission and discharge of pollution into air and water from factories, mines, storage tanks, septic systems, landfills, pipelines, wastewater treatment plants, recycling facilities, feedlots, vessels on the Great Lakes, construction sites, parking lots, and power plants. An active and healthy economy requires resource extraction and the release of waste. A building full of environmentalists working on permission slips for pollution is a dark kind of comedy.

At best—which it mostly is—the culture of the Whale can be described as mission driven. At worst, as a bunch of tattletales. Regulatory culture infuses every interaction. Use of the break room fridge requires the payment of dues, and a list taped to the refrigerator door lets everyone know who has and hasn't paid. Meetings need agendas sent out ahead in emails as well as handed out in hard copy. Making a phone call requires reserving a conference room a week ahead. Work plans are set two years in advance, and the state climate plan I am managing is not part of anyone's work plans. The improvisation that made me so effective in field science and the Senate has no place here. I don't know how to structure an agenda. Unscheduled calls from my large team bother cube neighbors. I love the bureaucracy's methodical approach, but the pace frustrates me. I itch to make a giant zigzag in black marker on the pristine bathroom wall.

One day I get a call from a staff person I have not met before. Within minutes he says, "You aren't from here, are you?" I ask how he knows. "You said *no* too fast." I want to protest that I am the girl next door. Friendly! And, if I were a man, would a *no* leave teeth marks? My cube neighbor invites me to use her compost bucket, and then uninvites me. My tea leaves mold in unacceptable ways.

A staff expert on climate refuses to work on my project. I push him by saying that his commissioner promised me he would. "Commissioners come and go," the expert says. It is true. The governor's appointed leadership will last only for the duration of the administration—or until an issue blows up. This expert has been at the agency since the 1980s and holds his own power. I too can earn such power. But addressing climate change needs to happen now. Besides, by the time I become structural to the institution, I might think every effort is futile, as this expert believes.

Media interest in climate makes my work flashy. Many state agency staff see the climate plan as a flash in the pan. Besides, public servants know better than to seek attention. Department of Agriculture staff line up like soldiers to defend farmers from any blame for their increasing greenhouse gas emissions. *Farmers are subject to acts of God* . . . Bringing up concerns about how we farm is like setting off a stink bomb in a middle school cafeteria. The governor has just resurrected my office from the ashes—it is yet to be determined whether it is a phoenix or an arsonist.

The trick of the public servant is to experiment wildly while staying invisible. Public servants move in increments so small that no one notices. It's the art of boiling a frog. A veil of boringness protects the bureaucracy.

Agencies defuse problems with preemptive engagement. We invite in the characters who pay attention and make their voices heard. Stakeholders range from chambers of commerce to nonprofits to the individual who publishes opinion pieces in the paper and shouts that the agency's deputy commissioner is a *whore*. The deputy does not react. Public servants don't talk back, even high-level ones. We read preapproved text from pieces of paper. *Only answer the question asked. Do not say anything unnecessary.*

My boss is my age and also came to the Midwest as an adult. He has adopted Minnesota-isms, mentioning the ice dams on his "ruff" and how he went fishing in the "crick." He refers to me as his pit bull. I do my best to meet his impossible timelines. He is hungry. I am hungry too. I have a baby and a toddler and want to get things done. "This is Anna, she is the brains," my boss introduces me to another manager. I appreciate my long leash. It means he values my work. But I miss the discourse and debate of my Senate office.

Leadership have their hands in so many buckets they can't pause, shine a light, and check out the details. Progress lithifies. My boss and the other two men leading the climate plan have other fires to put out. No time for discussions. They are gearing up for a political leap: to propose a law requiring that the state generate 40 percent of electricity from renewable sources like wind, solar, biomass, and small hydroelectric dams. But the data show that Minnesota's existing approved projects already have us on track to achieve that. No fight needed.

I call the three men and tell them we need to aim higher. They don't believe me. I wrestle everyone into conference rooms for leadership to hear the data from their own staff. Even after this, one of their commissioners continues to pitch the 40 percent standard. When I see this commissioner at a public meeting, he tells me he is hard at work on a state climate plan. I smile and nod. He is telling me about my own project. I am completely invisible to him. If the hierarchy were a piece of art, it would be a cubist painting of a beast with its mouth, mind, and body spread out across our forests, prairies, rivers, lakes, towns, and roads.

When the head of a legislative commission calls to invite me to present the climate plan, my boss says, "Pass it to me. I represent the work from our office." It is the same with calls from the press. "Relationships" aren't my "responsibility." We practice for my boss's moments in the limelight. I throw out question after question to ensure he fully downloads my knowledge.

The woman in the cubicle behind me has been in her spot for almost two decades. I wonder if that is my trajectory. Storybook plots have arcs, but in government life can plateau. I am afraid my enthusiasm and commitment to my job will shape-shift into a trap. I see how one gets stuck. I am firmly in my boss's shadow.

What will ultimately lead me out of the shadow first appears as a roadblock. As my work on climate change ramps up, with more staff participating, more public meetings, and the growing expectation of a legislative push on the proposed policies, my boss announces that our office has a water policy report due to the legislature in a few months. I couldn't know that the governor is about to put his full political capital into water advocacy.

The boss tells me and two colleagues to figure out the water report: content, structure, and management. *Just meet the deadline.* We need to amass a team from the staffs of eight state agencies who already have full workloads. State law lays out two pages of requirements for the report, some general and others hyperspecific.

A colleague from my office pulls me into the back stairwell, "We need to kill this report, the agencies don't want it." Another colleague asks me to walk so we can talk *safely.* Our building is cut off from downtown by highways and railroad tracks. We circle the concrete island, passing by the jail, pawnshops, and the Red Savoy, where pizza is protected from natural light by black trash bags taped over the windows.

Talking about water means talking about development, regulations, and agriculture—all open wounds. Staff have spent years in interstate and international negotiations to address dead zones in the Great Lakes, Lake Winnipeg, and the Gulf of Mexico. Minnesota holds all the headwaters. Working on the climate report, I learn that fertilizer, soil erosion, and animal manure, the major sources of agriculture's contributions to climate change, also pollute lakes, rivers, and drinking water. Department of Agriculture staff warn me, "Don't write anything in an email you wouldn't want to see in a headline in the newspaper."

Rocking my baby to sleep, I sketch out a plan for the water policy report. The next day I walk into my boss's office and tell him I will take the hot potato. I don't want to spend more time in stairwells. Maybe, just as much, once I formulate a vision, the idea tethers me.

I approach potential staff. "I'd really like to have you on my team," I say. This results in the same conversation over and over.

"What do you mean by water?" they ask.

"I mean water, like H_2O."

"I don't understand," they reply.

In government, water is defined by regulations: wastewater, drinking water, stormwater, water allocations for farms, trout stream water, groundwater, river water, lake water, impaired water, industrial water, waters of the United States, and public waterways.

"I will work on the report as long as it doesn't include wastewater," one staff person says.

"But that is your expertise."

"Exactly. I don't want the legislature coming after me." There are few rewards for taking risks, and the legislature scrutinizes any failure.

To bring on one staff expert, also a union leader, I use a card I have been holding. I become a full-share member of the union—*full share or fair share?* This seals not only his contribution to the report but also our friendship. I didn't initially opt to pay the full share because my position was temporary. My boss has just made me permanent, almost two years in, and only after I threatened to apply for other jobs.

I now better understand my role as a project manager. It means I walk in front. The report's outcome unclear, reception uncertain, I will be the one to trip over obstacles, walk into walls, and fall on my face. I will also draft or edit the written material and graphics.

The hierarchical system of state government, with its managers and appointed leaders, holds a paradox. Staff need authorization to bring up an idea, but they don't have access to their leaders. Leadership doesn't know enough about projects, or, sometimes, subject areas, to engage. Or they need authorization from someone above them, maybe even the governor. Many issues have consequences that are more immediate than those associated with the environment—that is an intrinsic challenge. Without direction from above, we freeze. We tiptoe. Accountability shoves us into corners. We take defensive positions.

Given the challenge of discussing water policy head-on, I step back to let the agencies react to each other. Where jurisdictions intersect is where I find cracks, black holes, blind spots, games of tug-of-war, and, sometimes, clinking swords. Pollution Control staff bring up how drainage from agriculture not only causes toxic algae blooms but also increases our vulnerability to catastrophic rainfall. Staff from the Department of Agriculture shove the spotlight onto the Department of Natural Resources, which allocates permits that pit the groundwater needs of wildlife against those of breweries, farms, and cities. Staff from the Department of Health raise the need to understand environmental problems as social problems. I listen, ask questions, and take notes. The tension between the thrill of experimenting, or going rogue, and the fear of getting in trouble fuels a tense and inspired report.

I don't anticipate the two things about this project that will get me off my career plateau. First, my boss knows better than to work on water—it makes the wrong kind of splash. When the report is done, *I* present the work to the legislature. I give talks around the state. Standing in front of crowds, I receive compliments. But mostly people put me on the spot. I have to answer to whatever they say.

At a rural community meeting, a man tells me that buyouts of flooded homes in Minnesota river towns might sound like a win-win on paper, but I haven't appreciated that when everyone moves away, the town dies. The community folds. I sit stupid in my car after the meeting. He is right. All solutions suck. Destruction comes in many forms, even protective measures destabilize us. When the rules change, jobs change, businesses fail. The iconic closing of coal-fired power plants threatens the tax bases of small towns. And yet we have to make changes. I can see that climate deniers aren't actually talking about science. The debate is about our soft belly fur: the need to protect ourselves.

The second thing to get me off the plateau: the governor is suddenly, dramatically struck by our widespread water pollution. He announces a controversial new regulation affecting farmers. He declares a day of water action and then a week. When he tells the press he will do a year of water action, his office advertises a full-time position for a water policy adviser. Once taken for granted, clean water is unbearable to lose. Water is life.

My boss emails me, "The Governor's office asked for names, send me a list by eleven." It hits a nerve. I walk over to his office. A young staffer sits across from him. "That thing you asked about, just give them my name," I say. My boss's eyes widen and then narrow. "We'll talk later," he says. The staffer looks from him to me. I flex my sweetest smile at them both.

That afternoon, as we're driving to a meeting, my boss says, "You can't go work for the governor. I need you here to write reports for me." I don't respond, and for me that is like shouting.

Through my own connections, I submit a résumé.

Before my interview, I hold Wonder Woman power poses in a bathroom stall. At the Environmental Quality Board, the governor's office floats opaque above us. The governor's staff organize meetings and show up an hour late for them. We expect to be made to wait. We hope and fear for the governor's attention. He could easily balance a budget by cutting our program.

In accepting the job, I become the governor's adviser, unofficially nicknamed the Water Czar. My last week before going to the State Capitol, my coworker and I spend our lunchtime meandering through the parking lots of the Whale's concrete island. "You made it," he says. I *feel* like I have made it. "It means a lot to all of us. One of our own. I can't believe it," he sounds genuinely happy. I had feared spending my life inside the gray Whale, but it has only been two and half years. This colleague and I cried in front of each other the day our kids started kindergarten. I am afraid I won't feel that comfortable with anyone at the Capitol.

When I leave, my boss confesses, "I worry people will think you are leaving because you don't like it here." He promises to throw me a going-away happy hour but doesn't. Instead, months later, he invites me to the office Christmas party.

During the legislative session my former boss calls often, and, generally, I choose to answer. "I am so jealous. You have all the power right now," he says one day.

"What power?" I ask.

I sit across from Governor Dayton at his dining room table and on a small plane where I pray to not throw up. I push Trump's EPA administrator on climate change and then offer to give him a tour of our State Capitol that he so admires. But working for the governor, I have no control over whether I make it home for dinner. I cannot promise my kids that I will be around on the weekend.

"You have all the inside information—that is power," my former boss says. I carry an invisible archive on my back with all the details left out of my concise briefings for the governor. The muscles in my neck and back ache from the weight.

At the State Capitol I see that even as the chief executive, the governor does not have the power to do what he wants. I've never seen anyone so disappointed in, and hopeful for, what the government can do. He expresses frustration at the constraints on his ability to improve the lives of Minnesotans: limited resources, divided government, bureaucratic processes, his age, personal conflicts. He cannot save the world alone or all at once. Actualizing change requires the collective. And it requires time—the key power that elected officials lack.

The governor might smack the conference table and cuss before meeting with his cabinet members, but he needs his leadership team. He needs legislators from the other party who control the House and Senate. He needs support from the business sector. He needs communities to care and make themselves heard. Building a coalition means compromise. Listening. Making trade-offs. Eating shit.

Some philosophies hold that life is a spiral, returning us to the same places over and over. But from inside, the walls and dead ends make it more like a labyrinth. Having power does not translate to feeling powerful. As I take my place at the head of conference tables and on stages, I can't shake the feeling of being a kid in the front row of class waving my hand in the air. I cringe seeing the twelve-year-old inside me who has to be faster and louder than her six siblings to tell a story at dinner. But isn't that what makes a leader? Enthusiasm and the willingness to play the fool?

Storming the Stage

At my first meeting with Governor Mark Dayton, I spill my tea when I catch the edge of the tablecloth as I sit down. This is not the entrance I imagined. I sop up the tea with a cloth napkin but can do nothing about the stain. The table hides my soggy suit pants. We are in the basement dining room of the Governor's Mansion. It is a world unto itself: Victorian architecture, quiet voices, staff who take your coat, geranium hand lotion, beautiful paintings, seasonal decorations, fine china, bottles of water in ice buckets, hot coffee in metal urns, and unlimited peanut M&Ms.

I've been hired to lead the governor's crusade for clean water. It's 2016, and I have two and a half years before the end of the administration. My first assignment: a water summit. The governor tells me it should be held in a rural community, accessible to everyone in the state. It should be an all-day event, with lunch provided, free to attend, and with speakers that engage and excite. I'll need to raise the money. "What might that look like?" he asks. I have four months to figure it out.

At my first meeting with the governor's environmental cabinet, one member jokes about sending another member's son a box of condoms at college. The cabinet is made up of the commissioners who head agencies with jurisdiction over environmental programs and regulations. Another commissioner points out the window at a colleague and says, "Looks like he took a shit in that hat." I can't tell when it is affection or animosity that fuels their roasting. I've been dropped into someone else's family. I sit at the head of the table with these six men who are fifteen to thirty years older than me, the only woman in the room. All of us white. They are recalcitrant Care Bears. Instead of rainbows and sunshine on their bellies, they carry images of mines, corn, and sewage treatment plants. I reach into the diplomacy tool kit I acquired growing up in chaotic households: deflection, avoidance, going along, and, most important, staying quiet. But while those might serve an immediate need, ultimately my job is to make these men take on and lead the governor's water agenda.

I try to make the cabinet members my teammates or friends or mentors. I bring fresh coffee and donuts. I explain the governor's goals and ask for their input. They let me know what they think of spending their time on yet another public event for the governor.

The governor tells me no one is allowed to yell at me. I should expect to be treated with respect. I laugh. Despite his well-intentioned declaration, what else can I do but laugh when a cabinet member tells me the governor should try doing his job for a day?

I know how I look to the environmental cabinet: stumbling and catching myself. Being a research scientist in politics is like walking with a steel-toed boot on one foot and a high heel on the other. I want to be a bridge, a two-way translator. I barge into strategy conversations to ask whether the proposed plan will be enough to prevent drinking water contamination. Science is all about asking questions. I pass out peer-reviewed papers to argue the case for explicitly stating that fertilizer in drinking water is not only potentially fatal to babies but also linked to health risks in adults. Despite my role in the governor's office, it feels like I am storming a party I wasn't invited to. *You again?* I am forgotten until I insist on discussing the governor's deadline.

The story goes that research scientists are pure, not in it for money or influence. They aren't leaders or decision makers. Scientists sit at the base of a pyramid generating the data and the framework that companies use to make products and political movements use to support policies. They aren't appointed to the governor's cabinet or the Public Utilities Commission. But this scientist left the lab. I put on a suit, and I want a voice.

The cabinet member who is most supportive, a team player

who tells stories about lemon bars in church basements that make grown men cry, tries to take me under his wing. He and his assistant commissioner, an older woman, sit me down: "Try to avoid talking, just listen." *Talk less, smile more.* But my goal is not assimilation. I am creating a world that is different from the one they inhabit.

When you try to storm the stage—step out, shout, call attention to yourself, be the change you want to see in the world—to make a difference, the reaction you hope to get is not guaranteed. It is still a question, an uncertainty: How can I most effectively exert my influence?

Is the All-White Male Panel Inevitable?

For the governor's water summit, I am told to ask cabinet members and allies in environmental and farm organizations for suggested speakers. We want inspiring leaders who will forward the governor's campaign for clean water. No surprises. The list I collect includes many familiar voices quoted in the media. When I show it to the governor, he asks, "Are these all white men?"

I didn't know we could ask that question. I am going to parrot it over and over—*Are these all white men?*

The governor's recognition that state government doesn't reflect who Minnesotans are came after he had appointed leadership in his first administration. The mainstream, well-resourced environmental movement has largely been urban and white, and this is reflected in who becomes an academic expert, a nonprofit leader, or a cabinet member. Trying to change course has been more successful in some areas than in others.

While the environmental movement is generally not run by research scientists, it carries that academic culture of viewing the environment as separate from humans. With campaigns, fundraisers, and policy, we can manage and protect landscapes and species. We fight for icons, like the image of a polar bear on a shrinking iceberg.

Senior staff in the governor's office ask why environmentalists don't focus on what is at stake for individuals and communities, such as public health and job opportunities. When I ask leaders in the state environmental agencies this, they tell me that is not their expertise.

The legislature reinforces the separation between issues by considering them in different committees. When we try to bridge a gap to protect well water, neither the Environment Committee nor the Health Committee sees it as theirs, but both have to vote for the bill to reach the Senate floor. The proposal never makes it out of the committee process.

The governor is not just asking cabinet members to include more people on staff, or onstage, or in stakeholder processes that inform government programs. He is asserting that we can no longer separate social issues from environmental issues.

Looking to broaden my list of potential summit speakers, I reach out to people across the state, researching and bringing names to the cabinet. "What about him? Or her?"

"I already gave you names," one cabinet member says.

"Anna, this is rural Minnesota," says another.

It's as though there is a big table with all kinds of people seated around it, but the light only shines on a few white men at one end. Pushing more and asking again, I get the light to shine on some white women (like me). But most of the table still sits in shadow. I see how the dominant narrative of experts and leadership writes a lot of people out of the story.

The State of the State

The governor's annual address kicks off the legislative session. I watch him projected on the wall of his Cabinet Room with his commissioners. The governor stands in front of Roman symbols of authority, fasces, carved into the dark wood of the House of Representatives.

On the screen, members of the Minnesota Legislature crowd

into the House chamber. The governor thanks his family and guests in the gallery above for coming. His eyes twinkle, and he waves up with enthusiasm at someone we can't see. Every chair is full, people stand in the back of the room. It looks hot. If possible, venues are kept icy cold for the governor. When I attend events with him, I wear long underwear beneath my suit.

I've fact-checked the speech several times and get déjà vu listening. Close to an hour in, he starts talking about dairy farmers Sherry and Vince. The health insurance plans available to them are prohibitively expensive, the governor says, "for a policy with a $13,000 deductible . . ." He takes a sip of water and looks out into the crowd. He begins again, "but-t-t-utzzzz." The word summersaults before landing, "but despite that awful cost." He looks down at the podium and then out into the crowd. He does not speak again.

A shiver runs through the governor's shoulders. A hyperlocalized earthquake is just beneath his feet. There's a slight delay between what we see and what we hear. As if to sound a warning, the smack of his head on the podium comes just before we see him collapse on the screen. It is a fall that cannot be stopped, just made gentler, and several hands lower him to the floor.

I don't want to watch, but turning away would force me to look at my colleagues. No one speaks. The chief of staff yelps. Ashen, she jumps up and runs out the door. I hold my breath. I've been in the office for only five months. *Could the governor have . . .*

A cabinet member yells, "Would somebody do something?! Don't you care?" She swats at the tears on her face.

"I could lead a prayer," offers the lemon bar commissioner. We put our hands in prayer positions. I close my eyes to hide from the room. This is not my religion, or my way of praying, but it feels good to be tender in a group that is anything but.

I don't know if this counts as prayer, but for a few quiet moments I focus on my favorite thing about the governor—watching him make decisions. He screws up his face, frowns, and looks like he is about to spit out a wad of gum. It takes physical effort for him to make the trade-offs: Fund health care for the undocumented? Or put that money into higher education? Why fund research institutions if elementary school kids aren't proficient in reading?

My shoulders unclench and air flows into me. A tear, slow as a snail, slides down my cheek.

Crack!

My eyes snap open. Bill, the governor's trusted adviser, stands in the doorway.

"You can go on praying if you like, but the governor just walked out of the Capitol on his own two legs," Bill says.

The next day, Tuesday, the governor holds a press conference. He announces that he has prostate cancer and takes questions. He cracks jokes at every opportunity, as if to say, *Do I look dead to you?* I've never seen him so jovial.

In the morning light, it's clear that what happened the day before was likely the result of his not eating more than M&Ms and coffee all day. Despite this and reassurances that his cancer isn't debilitating, we all get teary-eyed when our supervisor gathers us in her office. We have never seen her cry. It scares us.

On Thursday we go to the Cabinet Room to share chocolate cake and sing "Happy Birthday" to the governor. "This might be my last," he says. I choke on my cake. His tone may or may not indicate a joke.

The great truth is that a lot dies with us when we die. Another governor would not focus on water—my work is tied to this particular individual. We are paper boats, our ideas just steam rising off the Mississippi River. We sink, we fade, we are forgotten. The river never even realized we were there. It's as indifferent as time.

A photographer from the *Star Tribune* hands the governor a gift. There are no reporters at the event. Perhaps documenting the visual builds relationships, while probing with questions has a different effect.

The governor unwraps the present. It's a frame with two pictures. In one, the governor looks up and waves from the podium. In the other, his grandson waves back from the gallery. In that way, the story is rewritten. At least for him.

Drinking Wastewater Effluent

That Friday we travel 160 miles to the water summit in Morris, a windy college town close to the South Dakota border that advertises the impossibility of getting lost. On the campus of the University of Minnesota Morris, the governor stands in front of another packed room. He's casual in khakis and a plaid shirt with a suit jacket. Cheeks flushed, he looks rested.

"I want to let you know, we have some members of the working press from St. Paul who came all the way out here. I don't know if they came here to see my speech or to see if I'm going to pass out," he laughs. "But I'm not trying to talk for forty-five minutes so you'll be spared that."

In the afternoon, panel discussions spotlight the struggles and innovations of farmers, food companies, and local government leaders. During a Q&A, the chairman of the Upper Sioux Community, Kevin Jensvold, stands to speak from the back of the room: "Everything I heard today is about the dollars needed to implement the technology to create better health for our waters." There is an electricity to him. The room buzzes. The Upper Sioux Community reclaimed a portion of their ancestral homelands in the Minnesota River valley in the 1930s, lands they call Pezihutazizi Kapi (the place where they dig for yellow medicine).

Jensvold talks about how his community is investing in technology to reduce pollution far beyond what is required. *Money and regulations were not the primary driver.* He invites the room to drink with him, to taste the effluent from his community's membrane-batch-reactor sewage treatment plant. Drink sewage effluent? No one was expecting that. Standing in the back of the room, the chairman is center stage. He has everyone's attention.

The chairman was on our short list for speakers, but I'd never met him, and no one would give an endorsement. While registration for the summit was open to the public, we sent an invitation signed by the governor to each of the eleven tribal governments in Minnesota. An invitation from one sovereign nation to another.

The state has a long history of not being a good partner—and far worse—to tribes. More and more events I attend start with an acknowledgment of being on Indigenous land, but what this means or what should be done is not broached. Only during the planning process for the summit had I learned that this public liberal arts college started in the late 1800s as an American Indian boarding school. The full history of that school's treatment of its students is part of an ongoing investigation. About a quarter of UMN Morris students today are Native American and attend tuition-free.

The chairman's effluent is offered at the event's *Water Bar,* an art installation where visitors taste water from different places. *Guess which water is which.* The installation makes water, largely taken for granted, visible. A smiling woman presents me with four shot glasses of water on a carved wooden tray. "What does it taste like?" she asks.

"Minerally?" is untreated groundwater.

"Like your mouth when you wake up" is St. Paul tap water, my city's water.

"Chlorine" is treated Lake Superior water from Duluth.

"Lighter" is the chairman's community effluent. It tastes like *water.*

Giving Up Control of the Narrative

After the summit, the governor is lit. He announces water town halls across the state. His cabinet members roll their eyes at me. No one wants to talk more about water, they say, as if I could stop the train.

Sitting at the dining room table on the main floor of the Governor's Mansion, I broach my plan. "For the town halls, for the audience to see themselves in the speakers, to better reflect Minnesotans, I think that we should invite a broader range of people."

"Yes," the governor says.

"These speakers might have real grievances with you and the state. They might not have all nice things to say."

He shrugs.

"When we put them onstage, we might lose control of the message." I pause.

"Do it," he says as if we've had this conversation millions of times. But this is not just between him and me. I'm playing the conversation out for his leadership team around the table. I anticipate pushback when something unpredictable happens or an ally gets upset about not having a bigger role.

I take first steps. I don't know which are stumbles. I can't start from anywhere but where I am, from this inadequate place. I have not been given enough time. But I had better start. I work with a new member of our office who is a tribal member. Her presence opens new lines of communication.

For the town hall in the Twin Cities, I talk to groups representing different racial and ethnic communities and ask them to partner with us. The conversations are mostly my listening. Some want to partner. One group doesn't want its name on our flyers but requests to emcee a Q&A. I get our chief inclusion officer's backing to add a Q&A for this town hall, which might put cabinet members on the spot. My idea of experts expands from individuals engaged in water science or businesses linked to water to cultural leaders, community leaders, schoolteachers, elders, and youth. My role expands from translating between science and politics to building more human connection. Veering off from how things have been done, I risk making mistakes—I am certain to make mistakes.

At the governor's town hall in North Minneapolis, a historically African American neighborhood cut off from downtown by the highway, a commissioned art installation done by community members welcomes attendees. The NAACP has picked a youth to introduce the governor, a role usually reserved for mayors. I didn't ask them to vet their choice with our office, and they didn't offer. I meet the young man the night of the event. His activist history sets off alarms in the form of calls and texts to me. He captivates the room with poise and charisma. During the formal presentations, an elder weaves traditional stories about water. When she goes over her time, my phone vibrates, telling me to wrap it up so the governor can stay on schedule. I do nothing.

Despite the enormous loon puppet held up by anti-mining Girl Scouts, despite the teen protesters for whom the governor steps off the stage so they can take the microphone, despite it being the best meeting ever, the Twin Cities event doesn't register in the mainstream media. By giving up some control, as in stepping back, as in letting a message emerge organically, as in people sharing their relationship to water in their own way, we have produced a story that the media doesn't recognize. We get no headlines.

Small Acts, like Ice Cream Cake

It is the last night of the legislative session, and we're waiting for the legislature to make the next move on budget and policy negotiations. Anything could happen. We are on guard and anxious. The work of months, or, in some cases, years, will be determined in the next few hours. Silly-tired with the sour smell of adrenaline, a colleague and I run out of the building. We laugh like kids skipping school and get to Dairy Queen just before it closes at 10:00 p.m. Agency staff and cabinet members watch the legislature's proceedings in a Capitol meeting room. We bring them ice cream cake.

They eat less than half, and we bring the rest back to our office. A colleague who DJs on the side turns on music. *If you need me, call me; no matter where you are, no matter how far.* Staff shimmy into the room, shaking hips and shoulders. We hand out slices of cake. This will be the story we tell after the legislative session is over, the one that makes us feel like part of something special. The one that makes us smile after we laugh. *I told you; you could always count on me, darling.*

Someone yells, "Take cake to the governor!" We fall into a conga line, the DJ keeps playing, *Here I am baby. Signed, sealed, delivered, I'm yours!* The rustling of my colleagues' suits and the clicking of their shoes on marble let me know that we are in this together. The governor, he is alone. One is never more alone than at the top. But we are going to storm his office.

Protesters are camped out in the reception area, so instead

of walking directly across, we go down three flights to the basement. We cut through the security room with its video monitors. "Want cake?" we offer the familiar faces who drive the governor and stand in the shadows at events. We grab the chief of staff as we dance our way up three floors.

The governor opens his door, happy to see us. *You know your love, keep on lifting me, higher, higher, and higher. I said your love, keep on lifting me, higher and higher.* He smiles at the half-melted cake. It is this smile that makes us want to dig deeper, to find energy we don't have. To let him storm through our lives, take away time with our families and weekend dates with friends. It is what keeps us believing in him even when he confounds us.

He gets enough chairs for everyone to sit, but we don't. It starts with a sway, a bounce, a smile—arms lifted by angels. The music is in our feet, our hips, our shoulders. We eat cake with our hands. This is my second or third slice, the contrast of smooth ice cream and cookie crumbles more perfect with each bite. *You can dust it off and try again. Try again. Cause if at first you don't succeed.*

He dictates a text, his body woman types, telling the Speaker and Majority Leader, both from the other party, he is headed home for the night. It's past midnight. He says, "Don't hesitate to call, I will answer." These are things that no one sees: the governor urging his opponents to wake him, him finding chairs for us, our shared laughter. This is what takes place outside the performance. Not everything makes headlines. Sometimes what is important are a million small interactions, *don't hesitate to call, I will answer.*

I want a story of bold change. One where we hold policy accountable to science, where the government reflects people in all forms of diversity, where environmental advocacy includes public health as key to the story. I want to throw open the door and let in the breeze we've been waiting for, the wind that is going to lift us all up. But change can only take place one step at a time, one brick laid, one person listened to in one moment in time. And that is too slow.

Strangers Form Constellations
That Make Sense of the World

I lock eyes with passersby, call out to bring them to my table spread with blank books. Children in bright wool and boots, ironic college students in leather and tutus, men with wooden coins in stretched earlobes, a mother with a full sleeve of sailboat tattoos, stooped couples with gray hair holding hands—everyone has vinyl records in their arms. The Record Store Day street festival is a yearly event with beer, live music, pottery for sale, and now the Little Poetry Library. I hope no one from work spies me. My work phone is turned off. I am stealing back a weekend for myself.

I've kept my art entirely separate from my professional career. Advising elected officials, working for the governor's administration, I have to keep a low profile. But I am still a writer. Just as I have always done, I write in my journal or use the back of a notebook to render the drama of a closed-door meeting. Sometimes, I publish under a pen name. Setting up a booth at a street fair today is a way to participate in the broader community as myself, with no agenda beyond poetry.

"How much does it cost to write a poem?" a redheaded teen asks. She is with a friend who has similarly long and shiny hair.

"Nothing," I spread my arms wide. At the office we work under pressure to get things done. But here, I bask in the glow of an entirely optional activity. Nothing need happen today.

The teens look down at the colorful books on the table. The smallest book is one inch by one inch, the largest the size of a

Hallmark card. They choose one with a cover made from an article on kefir and the words "Wild Season," cut from a Walker Art Museum flyer. They walk away with the book and two pens. I wonder if they will return.

Poetry can be a methodical process, or it can be blowing words around, like dandelion seeds, to see where they land. Either way, it is an experiment. The search for knowledge through art can be playful, technical, weird, and sometimes raw. The effort it takes to carve out time for art in my own life makes me want to share that space with others.

The Little Poetry Library sits at a corner on Lake Street near the Mississippi River in Minneapolis. It is like a small dining room cabinet set on a post about four feet tall. Names of poets cascade in appliqué, with a girl walking into them as though she is entering a forest: Alice Walker, Walt Whitman, Naomi Littlebear, Marge Piercy, Gloria Anzaldúa, Louise Gluck. "Where do our words come from?" asks a cloud overlapping a blank library card. The city has dozens of free libraries put up by community members to share books, but this is the only one dedicated to poetry. It sits in front of Hymie's Record Store and the Blue Moon Café. Inside the library cabinet, visitors can sign a notebook and pick from a handful of skinny books as dense as cattails in a marsh. More books go out than come in, as if people only realize their thirst when they open the library door.

I made the blank books that are now spread across the table using an empty baking soda box, recycled wrapping paper, a 1960s sewing pattern, and Nikki McClure calendars. Like a bird perpetually building a nest, I fill a desk drawer with paper treasures. For today, the book pages are a mix of white paper and old USGS maps from my PhD field sites. I stapled the spines and then smoothed colored duct tape on top.

I met with the local poet who built the library, Carolyn, in a living room with children and cats climbing in our laps. We had become friends in a fellowship program a few years earlier. We plotted advertising and table layout, wrote prompts on index cards. But the actual doing of it, the soliciting of strangers, only

I was excited about. My vision was clear. I would rub writing prompts together to spark the fire of poetry in strangers. A current of looping script on blank pages would continuously pull the crowd to my table. But once the table is set, I feel the discomfort of an oversize sales grin on my face.

A man in red flannel with a big beard and weather-worn cheeks stops at the table. He chooses a book with a tree on the cover—one made from an outdated calendar. He goes away and sits at a café table. About fifteen minutes later he comes back, hands the book to me, and waits as if I am the teacher who will grade it.

"Please, leave it in the library," I say, and lead him to the small box. "You are an author now."

He blushes.

After he is gone, I pull his book from the library and read:

Dog song fleeting
With wolf songs
Wily Coyote
Dancing to hills

I hold out the basket of prompts on index cards to a mother with a sleeping baby. *I woke up to the sounds of* . . . she reads out loud before writing, "crying, babbling, laughing, tears falling, and squeals."

An older woman with hollow cheeks and the symmetrical features of a doll pushes away the prompts. She tells me, "I have a lot to say." She returns an hour later, her book, twelve inches by twelve inches, filled with poems, each page covered in tiny and neat script:

Birds—
Were a talisman metaphor for my mother—
When I was young she would lament my sure earth bound
Leaps from the nest—
In later years her own wings clipped she embroidered them on
     muslin sheaths—
All those years later they are still on my green walls.

I have a basket of objects to inspire poems: a glow-in-the-dark bat, a pine cone, a gelatinous blue lizard, a shell, a ceramic crane, dice, knitting needles, a wooden block with the letter W on it, a small brown book with Saint Francis's *Cantico delle creature*. I promised my husband I would bring the playing cards back home—three of diamonds, ace of clubs, jack of hearts—but a man with strong shoulders and torn jeans takes the ace and writes:

The windmills in Oklahoma point in all directions

The card becomes a book in the library, something that can be taken and returned or just taken.

A middle-aged woman with long gray hair tucks a book into the library, I pull it out a few minutes later. The two-inch-by-two-inch book is made from an old cat food bag.

Also great are pine cones and playing cards
They help people make important decisions
. . . Knitting needles are great, they help things get knit
That is what you want

Some poems are instructional, letters to the self sent outward:

Lose yourself, just let go
No matter what your fourth grade teacher told you
Poetry has no rules.

This poem is titled "Last One, Really." It is the sixth in a series by the same author, but only at the halfway point of the book. I offer the rest of the book to a woman who says she is intimidated by the blank page. She has a grocery bag from the new market across the street in one hand. I peek when she returns the book. It now reads as a conversation between poems:

So is this a poem, who knows?
I don't
People squinting without sunglasses, yet it isn't really a sunny day

A retired librarian with a helmet of gray-blue curls popping out from a black beret tells me, "I love this idea."
"Do you want to write a poem?" I ask.

She waves her hand at me, as if to say, *Get outta here sister,* and walks away shaking her head.

"Hi, you want to write a poem?" I call out. A man in leather with a Mohawk shakes his head. A mother in a red trench coat rolls her eyes. A thick forty-year-old man in black with an armful of records says, "I will think about it." A sinewy man in his early twenties with ripped jeans says, "I will walk around and look for ideas."

I want to call after them, "Don't stop to think." I want to yell that when we write on the open page we make something that has never existed before. We all carry the weight of two hypotheses when we start something new, whether it is art or science or even a political campaign:

H1: This is an interesting idea.
H2: I am not crazy.

Two men stand in front of my booth holding out a book. One has shaggy hair and dimples, the other wears a leather jacket over a thermal undershirt.

"You do this often?" Dimples asks.

"Ummmm, no," I laugh and look up.

"Is there a rule against standing here?" Leather Jacket asks with a serious voice and follow-up giggle.

I look up. They are cute. Suddenly the live music behind me thuds against the pavement. I think of running into the current of people and beer. Instead, I blush and concentrate on arranging books and pens. They giggle, cough, clear their throats. I hunch my shoulders and don't look up again until they leave.

The book they set in front of me has a crescent moon on the cover. I open it and read,

> . . . that split second moment of connection with a complete
>   stranger across the room.
> Too real
> Too intense
> Too scary
> Look away

I hadn't considered that I too am a prompt.

Passing as young and carefree makes me miss my kids. I pick up my phone and tap out, "How is naptime?" A minute later my pocket vibrates, "Mac'n cheese for lunch. Both boys asleep. House is peaceful. XOXO." For me, mother guilt comes in the form of doubt over doing anything not for or with my children when I am not at work.

I like watching from behind the table. Dads with preteen girls wait happily while the daughters fill page after page. The dads don't ask to be included, and they are in no rush. They have seen the future and are thankful to put anything between them and the passing of time. Parents with young children groan as if this is another task, another test where they must rein in their wild animal. Many children just want a book to take. The books *are* pretty. A group of teens in layers of backpacks, denim jackets, flannel, spring dresses, and hoodies laugh and write. One by one they put their books in the library. They do not share their words with each other.

The books come back faster than I can keep up. I lose track of who wrote which. The poems are now mysteries I cannot connect to individuals.

Some poems rhyme:

No one knows I've got golden toes
Polished and painted and fine
In holey socks and low-soled shoes
I almost feel them shine. . . .

Others list:

Bike
Bike
Bike
Coffee

The April wind picks up. I rush to cover the books from raindrops. My tent shifts in a gust. The woman selling pottery at the table next to mine grabs a tent pole, and together we brace against the wind. "Would you like a book to write in?" I ask. Quarter-size marks on the pavement make *plop plop* sounds, the mineral and dirt smell of rain fills the air.

"Oh, I wrote a book already. I keep watching people's reactions when they read it," she laughs. Her long curls shine with droplets of rain.

Did she write?

Roses are red
Violets are blue
Sometimes I fart on escalators

The drops fall faster and harder. I hurry to pack up the handful of blank books left on the table. A bluegrass band is playing, and the crowd in the street behind us shouts and calls out. They are immune to rain after a long winter. I peek in the library one more time. I pull out a book two inches long and one inch tall made of an old cracker box with a cover that reads, "Stone Ground Crackers." It has just one word per page:

I
did
it
I
really
had
an
abortion

I look for signs of relief or sadness or laughter in the script of the small book and then I place it back on the shelf for the next reader. My daily life doesn't include space to hold the paradoxes of life, never mind let emotions surface. The poem opens a door for me, perhaps it will for others. Reading about someone else's experience can stir delight or rage. It might make me cry. Art connects me to both my most spiritual and my most animal selves. Sometimes to see myself, I need to see others. As in, to know myself fully, I need to live in community.

The Irrational Logic of Our Bodies on the Landscape

"I pledge allegiance to the flag," I say in a slow-motion stage voice. Three hundred people stand up and join in, *and to the Republic for which it stands* . . . I haven't said these words since high school, and then I only sometimes stood. But water meetings in Minnesota start with the pledge, and today in St. Cloud, I am the master of ceremonies. The pledge is not recited at meetings on workforce development or railroads. Just water. It asserts the fundamental covenant that the government provide safe water to the people.

The red tile floor of the St. Cloud Technical & Community College cafeteria is barely visible between the plastic tables with seven or eight people crowded round. Our office has evolved a shared shorthand to operate at full speed. We refer to the governor as GMD. Tonight is our halfway mark in traveling across the state to engage the public in discussions about how to improve water quality. An undercurrent of anxiety pulls at me. Sleeping in roadside hotels, I miss my small children.

GMD planned to attend only the first town hall out of the ten, but he has come to them all. By that measure, as well as the front-page stories in rural papers, the hundreds of strangers shaking hands, the inspiring stories, and the healthy debate, the town halls are a success. But our allies and collaborators—environmental nonprofits ("the Enviros"), philanthropic organizations, and state government leaders—are skeptical. They pose a warning and a question: Can people talking to each other really achieve anything?

Agriculture is responsible for 90 percent of the pollution in the Mississippi River, which flows less than a mile east of this cafeteria. That pollution contributes to a dead zone in the Gulf of Mexico comparable in size to New Jersey or Rhode Island, depending on the year. Some 40 percent of Minnesota's lakes are not safe for fishing or swimming. At least 10 percent of families with private wells drink water with enough fertilizer in it to be fatal to infants and potentially cause cancer in adults. Toxic algae kills dogs that drink from polluted lakes in summer. There is metaphorical and literal shit in our water. All this in the land of sky-blue waters, the land of ten thousand lakes, the land where we pledge allegiance to our country before we talk about water.

In *Braiding Sweetgrass,* Dr. Robin Wall Kimmerer challenges us to see our relationship with nature as reciprocal. I was trained as a scientist to see nature as the subject: indifferent. To do otherwise was anthropomorphism. When I read Kimmerer, I know with my body, as much as my rational mind, that there is a depth beyond admiration when I step off city streets. For all that I learned modeling aquatic ecosystems and calculating changes in water chemistry, science did not teach me to understand people as part of environmental systems. I struggle with the question: Does nature love me back?

The governor's environmental concerns came during his second term in office, when he read a headline about southwest Minnesota having no swimmable lakes. Outraged, he dropped the "Buffer Bomb" in 2015. Farms are largely exempt from the federal Clean Water Act. The governor's new regulation required perennial vegetation between waterways and row crops to prevent agricultural chemicals from flowing directly into rivers, streams, and lakes.

The regulation was punctuation in a conversation that had made no progress for decades. A top-down command. But the governor had no initial discussion, vetting, or negotiations with interest groups. The Enviros took great offense. They found out about the Buffer Bomb along with the public. They felt it was their right to shape and lead any environmental law. They are still furious. Insulted. And the mandated buffers reduce only a fraction of the pollution. Some would rather have no regulation than an imperfect one.

Driving to town hall meetings across Minnesota, I see the miles of grass and wildflowers filtering out farm chemicals along streams and rivers. It is thrilling. Over the past two years, the legislature has attacked and taken small bites out of the buffer law. These ribbons of green are evidence of my daily work at the Capitol.

Nicknamed the "Buffer Bully" in the media, the governor saw the next step as building a water ethic—a bottom-up approach. Get people to talk about water. Talking ignites action. *It started with a headline.* While I am leading his town halls and campaign, I am not sure what a water ethic looks like. No matter our individual level of care, the system in which we live, work, and play creates pollution. I don't see the logic. Without bigger, more systemic change, are our individual actions meaningful?

After the pledge I sit to the side, where I am out of the way but still there—*just in case.* Only six months ago the governor fainted on live television. Applause rings through the room as the governor steps onstage. He wears khakis and a maroon polo shirt with a suit jacket. His asymmetric smile makes everyone forget he is losing and catching his balance as he walks across the stage.

He opens his mouth to the brightly lit room. Breathes in the crowd. "Welcome—" That is when they charge.

As if on cue, as if this were a dance performance, a group of protesters burst through the door yelling into megaphones. They form a line clad in jeans and sweatshirts. One wears glasses, another a bandanna. They take off overstuffed backpacks and place them on the floor.

I'm hyperaware of my body with its kitten heels, blue dress, stud earrings. This is all that is between the protesters and the governor. I fill my lungs, ready.

The protesters talk about Enbridge's proposed Line 3 pipeline carrying Canadian tar sand oil from Alberta to Wisconsin. They speak of our need to stop building fossil fuel infrastructure that locks us into irreversible changes in climate and the inevitable spills and leaks. It doesn't matter that I agree with them. It's their overstuffed backpacks, not their words, that unsettle me.

My phone vibrates, "Tell Arnold to stand down," a message from the governor's assistant. I realize a bodyguard is beside me, his bulk moved into place with the grace of a ballerina. I hand him my phone to see the message. At the first big water event with a thousand attendees, protesters coordinated with security. *We'll charge from stage left. Would you mind coming from stage right? No problem. Feel free to stay for lunch* (they did). But these protesters are not known to us, not affiliated with an organization.

The governor protested in his youth. I can hear his voice in my head: *Give them the microphone, let them be heard, if they won't leave after they have said their piece, end the meeting. Don't fight for control, don't fight.* This year the legislature passed a bill to increase the charges for protesting from a misdemeanor to a gross misdemeanor—a direct response to Black Lives Matter demonstrations. The governor vetoed it. This was not simple, as the bill also contained a provision allowing undocumented immigrants to get drivers licenses. The bill was like a sandwich made with fresh bread and shit in between. A classic political move: *the poison pill.*

I look out at the sea of faces, more old than young. The room is waiting. The protesters aim their speeches at the soda machines against the back wall, the closest thing to decoration in the cafeteria. They are in tennis stance, ready for balls to come flying from any direction. The governor clasps the podium with both hands for balance. I recalculate the meeting schedule. The goal is for the crowd to be heard. My biggest worry has been that I might not hear what I don't expect.

"Booooo!" shouts someone from the front of the room.

"Go home!" comes from the back of the sea of tables.

The governor shushes the crowd, "It's their First Amendment right to speak. Let them say what they have to say."

One protester hands the megaphone to the next. She begins a crisp call to action, touching on treaty rights, the headwaters of the Mississippi River, and farmlands. She passes the megaphone to the next person, who says pipelines are icons of inaction. Like dominoes smacking, they pick up one after the other.

I too want clean air and clean water. Under what circumstances would our roles be reversed? Why am I in heels and a suit jacket? Is it a difference of personality? Am I giving less? Do I need to be louder?

Being part of the system makes me crass about contradictions. Wins have narrow margins—just squeeze through. We have divided government, both politically and geographically. Political parties in Minnesota straddle mining communities up north and urban professionals in the cities. The parties believe they have no choice but to say *yes, and.* While sentiment is strong for action on climate change and environmental justice, it is also there for pipelines and copper nickel mines that make climate change and pollution inevitable. *Yes, and.*

In the crowd, I recognize the leaders of farm groups, each sitting near an exit. I think of how they know the environment through the muscle memory of plowing fields. Ahead of announcing the town halls, the commissioner of agriculture and I called each of them so they would know our goal was not to announce a Buffer Bomb sequel. It took three hours. The commissioner joked, "It could be worse, we could have diarrhea" and "If you want loyalty, get a dog." The farm groups stayed neutral on our budget request to the legislature for the town halls. They helped us find farmers to speak about clean water practices. Their help was the equivalent of finding a rose in your high school locker from someone you didn't think even liked you.

Conspicuously missing from the town halls are the urban Enviros who work to protect water and natural landscapes. Early in the planning, we invited them to tell us what they hoped for in the town halls. They said, "You will achieve nothing by going out and listening to people." The philanthropic organization that had previously funded the governor's public engagement declined this time, given the Enviros' boycott. This was the equivalent of your steady sweetheart breaking up with you in a YouTube video.

In *The Organic Machine*, Dr. Richard White says, "One of the great shortcomings—intellectual and political—of modern environmentalism is its failure to grasp how human beings have historically known nature through work. . . . Environmentalists stress the eye over the hand." Media headlines focus on urban/rural divides, but I suspect the divide is less about where we live and more about how directly our income is tied to the land.

The governor's cabinet expressed a similar sentiment to the Enviros. *We already know what we need to do. We won't make progress sitting around talking.* It was with rank-and-file staff and local governments that I found support and collaboration. The cynical part of me thinks hope is tied to hierarchy. As in, the higher one gets, the less one believes in change. Or the more invested one is in the status quo. My practical side knows agencies are stretched too thin and challenged to take on the new workload without notice.

An urban legislator left me a howling voicemail about a farmer who spoke at a town hall. "He said MOTHER NATURE will heal everything. HOW COULD YOU PUT PEOPLE LIKE THAT ON STAGE? Why didn't you have university scientists and REAL EXPERTS?" For each town hall, I form a local advisory group. Based on conversations with them and others, I invite speakers from the local community: farmers, high school teachers, public health workers, mayors, tribal chairs, and local government staff. Some speakers are captivating. Some express great distrust in government. Some have bad blood with state agencies. Some likely don't share our political party. Some fumble in search of the right words. I cross my fingers when they take the microphone. I am not always sure how to integrate their lived experiences with scientific data. Regardless, I listen.

After all of the protesters have spoken, they fall silent. They stand still, blinking and catching their breath, the laser from the projector in their eyes. I realize they might not have scripted an ending.

The governor addresses them in a neutral tone, "You've had your chance to speak, will you now let everyone else have their chance?" The room watches to see what will push him over the edge. I'm not worried about his temper. I'm worried about how long he has been standing at the podium.

The governor gazes out at the room. The protesters look at each other, to the crowd, and back at each other. One talks into the megaphone, but this time he rambles. His voice falters. Once you have control, do you give it up?

Someone in the crowd claps. More people join in. Palms turned to drums, tables of farmers, local government staff, environmentalists, engineers, grandparents, schoolteachers, deer hunters, retired mine workers, union members, cabin owners, and college students release their clenched shoulders and the grim thoughts that shot through them with the takeover. We each have a right to speak, but there is no requirement that anyone listen.

The protesters try to yell over the applause. They look at each other, resume tennis stance. It is hard to know the line between pushing far enough and pushing too far. The protesters look at the clapping crowd, look at each other. And then they fold. They walk out to the slap of water on rocks, the three hundred pairs of hands here for the process. It's a *Minnesota nice* fuck you. Democracy overlaid on democracy.

What falls out of the protesters' backpacks is a love potion. The surprise of their arrival, and the relief of their departure, makes us look around. We are grateful for the dingy room, for each other. We didn't know if we were safe, now we do. We smile bigger, talk louder, laugh quicker. Like Popeye with his spinach, we guzzle their youthful spirit.

A dairy farmer takes the podium, a public works director bounces across the stage. With each speaker, a knot inside me loosens. Finally, our destination, what all this was meant to prepare us for. I stand at the podium, "The first discussion question . . ." Three questions will lead the room through a logical conversation on what people value about their lakes, rivers, and drinking water, what they want to see improved, and how they want that achieved. At the end of the town halls I want a to-do list: a plan of possible legislative changes and meetings to schedule.

I pull up a chair at the table with the governor and an assorted group. He has told me, "The day I ask for talking points is the day you can take me out back and shoot me." But I still need to take notes on what he says.

On each table there are giant pieces of paper and markers, instructions for a phone app to enter discussion points, and blue spherical pieces of paper for sharing individual thoughts. The blue spheres are meant to be water drops but could just as easily be tears.

"What you did with the buffer regulation," the woman looks into the governor's face. He squints at her and leans forward to hear. We all lean in, elbows on the table.

"It was the right thing," she says.

I smile, but then I feel the itch. I want to ask her, *Was the Buffer Bomb worth losing the House and Senate?* I've heard from more than a handful of legislators that our party's poor outcome in the last election was a response to an urban governor regulating rural communities. It is unbearable to consider what we could do if we didn't have divided government.

"Buffers aren't about protecting water, they are a government taking of my property!" an older man at the table says.

"My family drinks bottled water. It's just what we have to do in farm country," a younger man with a baseball cap says.

"What does someone sitting in an office building in St. Paul know about farming my land?" a middle-aged woman snaps.

One truth completes another. Farm bankruptcies are on the rise—as are farmer suicides. It is a trade-off: remove acres of crops and potential profits to reduce pollution. There is evidence that over time clean water practices can lead to higher yields and increased profits. But farmers depend on loans, and neither local banks nor Wall Street's secondary markets are interested in long-term outcomes. I doubt that banks even realize how they institutionalize the poisoning of our lakes and rivers. And, in our democracy, nature doesn't get a vote. Future generations don't have a voice.

When I started working for the governor, I spoke with Greg

Page, former CEO of Minnesota's food giant Cargill, the largest privately held corporation in the United States. We met at a grocery store café near his home. He wrote out an equation on a sticky note—his take on the Buffer Bomb:

$$\text{Dissatisfaction} * \text{Vision} * \text{Quality of the First Step} > \text{Resistance/Trust}$$

$$D * V * FS > R/T$$

Simply put, the desire for change has to be greater than the resistance to change. And nothing happens without trust. That didn't bode well for the buffer regulation because most people aren't aware of water pollution. It often has no taste or smell (i.e., dissatisfaction was low). The governor's announcing the buffer regulation without talking to interest groups felt like an attack on rural communities and the Enviros (i.e., low points for the first step). And no one trusts government (i.e., resistance was high). Maybe any new regulation is not so different from how, despite my agreeing with the protesters, I want them to leave—maybe whenever someone else forces punctuation into our lives, our impulse is to punch back.

I check the phone app to see what other tables are uploading from their discussions: *Get rid of the layers of government, change requirements for manure management, for the good of others and not just ourselves.*

My phone alarm goes off. "Time for the next question," I whisper to the governor.

"We are having a good conversation," he protests. "I am interested in this dairy farmer."

"I have the schedule all worked out," I assure him. He looks as though I turned off a movie right before the climax.

At the podium I clap cymbals together. The shrill ring reverberates through the room. "For the second question, please go to a different table with new people." The room unfolds as everyone stands and makes way for one another. The governor moves over a few tables.

I jump off the stage, look up, and realize I'm in front of two legislators. I shake their hands and thank them for coming, a smile plastered on my face. They voted against funding these town halls. "It would be letting the camel's nose under the tent," the Environmental Committee chair had told me. It's a common saying at the Capitol, but since the chair threw it at me, the phrase has rolled around in my head. The committee chair doesn't see any harm in the town hall meetings, but in divided government, he doesn't want to be seen working with us.

In what feels like a few seconds, we move to new tables for the third question, cabinet members make closing remarks, and then it is over.

"Thank you," a large man comes up to me. My hand disappears into his.

My phone pings, it is the governor's body woman. "Come outside." I take my leave from the man before hearing about his appreciation for the meeting or his anger at government.

Fresh air feels good. I am amped up. Steam rises off my suit jacket. The governor and members of his cabinet huddle with reporters. His assistant intercepts me, "That group over there. Can you talk to them? GMD has an early morning and it's a long drive back."

I walk up to an older man and woman. "I'm the governor's adviser. I heard you had some concerns?"

"The county won't listen. . . . If your office told them to do something, they would." The woman's eyes are big, her hands wave to emphasize her words.

"So you have a dispute about the land by the lake being part of your property?" I ask. The couple are emotional, and until they get their anger out, the details gum up. If this is the county's jurisdiction, the main service I can provide is listening. At every town hall people come to me with stories of county, state, and federal permits—none coordinated, not even within the same level of government—that sound like allegories of bell jars and spider webs.

I watch the governor and his body woman walk to the black SUV, get in, and drive away. I let out a sigh. My shoulders relax.

Back inside, I hear my name, "Anna!" I turn and see an acquaintance who works in corporate sustainability. His perpetually flushed cheeks and blond cowlick make him look freshly roughed up by the wind.

"Did that make you hopeful?" I ask.

"People came. They talked to each other," he smiles.

"But what does talking translate into?" I ask.

"The point is that people were here," he says.

"Is that enough?" I feel the prickles of fatigue. My adrenaline running dry.

"It means something," he smiles.

The room empties. A vacuum cleaner roars. We keep talking. It feels good to not just listen, but to exchange. I am no longer hosting.

At a bar we hash out our ideas. We want outcomes as crisp as our nachos, as refreshing as our beer. It's past midnight. Possibility pours into me. I don't want to go to sleep—I want to run a marathon.

* * *

"Do you remember the St. Cloud protest?" I ask a few months later.

The governor wears a flannel shirt, and one of his socks has a hole in it. The white tablecloth of his dining room is littered with M&M wrappers and Diet Coke cans. There are homemade cookies on a gold-rimmed plate, and the sun streams in a window overlooking the garden with its fountain.

"I was ready," I tell him.

"Ready for what?" he asks.

I open my mouth, but nothing comes out. I know that he is fatalistic, if someone wants to attack him, they will. I close my mouth and shrug.

"What would I have told your sons if something happened to you?" The tone of his voice says, *Are you fucking kidding? My life is not more important than yours.*

I was prepared to protect him—from fainting, from protesters, from inefficiency, from inconvenience.

"I really enjoyed talking to all the people at the town halls," he smiles.

A strange unease comes over me. "Enjoy" is not a word I would use for the town halls. I was counting the minutes: delivering a show. But for him, it was always about the people, in any and all forms. That was why he traveled the state. He was focused on making connections, not on achieving tangible outcomes.

I think back to the ten times I raised my voice to pledge allegiance to our country and the hundreds of voices that joined at each town hall. I didn't stand for the pledge in high school because I felt like the god it referenced was not mine. But what links my homeroom in Providence, Rhode Island, to the water town halls in rural Minnesota is less about the words or what they mean than it is about the power of voices joining together. We connect with each other when we do the same thing at the same time.

At the town halls, an old man talked about farming his family's homesteads. A tribal chairman presented on stage and spoke so quietly it forced stillness over the crowd. A group of older people in matching blue pro-mining shirts threatened to cause a disturbance but then dissolved into discussion tables. A Somali member of a school board in southern Minnesota filled a table with his daughters. All this I hold on to. I treasure. But when the change happened inside me, I was standing outside a bathroom in North Central Minnesota with a group of protesters who, just a few minutes before, in the main meeting space, had shoved phone cameras in my face and shouted at me. In the hallway, we circled up and spoke of being mothers. One of the protesters had a white plastic mask pushed up on her forehead. That was when we looked each other in the eye. It resolved nothing concrete. We were just a circle of mothers seeing each other's beating hearts. Perhaps being in relationship with nature starts with relating to each other.

From these town halls I hold on to truths that go beyond science. Science in our lives is part of larger narratives. And what we believe in are our stories, not science. We live by our stories.

V. Frontierior

Frontier

1   a border
> The philosophical *frontier* beyond which science has
> no answers.

2   a new field for developmental activity
> Every tenth of a degree of warming will lead to
> increased deaths. Going beyond 12°C warming is a
> *frontier* to an uninhabitable planet.

Interior

1   belonging to mental or spiritual life
> A vibrant *interior* not reflected in my generic skirt suit.

2   lying away or remote from the border or shore
> The chants and songs of protesters carried through
> the marble floors and walls to my office in the *interior*
> of the State Capitol.

Frontierior

1   the internal experience of a life transition
2   a cross section of breaking apart and healing

Not Funny

♦

Reckoning with Complicated Truths

I am in a cabinet meeting on sexual harassment policies in Minnesota when the news breaks—any humor in that irony is lost to the gut punch. It's November 2017, and #MeToo is on every front page. Power is reversing course and flowing in a different direction—at least for the moment. At least for a few women. The news about my former boss, Senator Franken, arrives in a text from someone sitting across the table. It is a text of a tweet of a photo, an image from his previous life as a comedian. I can't make sense of what I see: hands hover over the chest of a sleeping woman. A slice of a joke that has time traveled and is now everywhere on social media.

I look around for an indication of whether others have seen the news. The colleague who texted opens her eyes wide like a wrecking ball is coming at the conference table. I know from working in politics not to act on impulse when emotions run high. I need to know where the photo was taken. Who else was there? Why?

I've had many bosses, all of whom I learned from, many I liked, but this boss set me on a path with an actual career trajectory. And he hired me six months pregnant. He wanted to change the world and believed I could too. My bosses have more often seen their staffs as resources to extract, tools to make use of, or entertainment for the ride. My professional growth came from pushing against them. With this boss I was part of a team.

I nursed my baby in the senator's bedroom during a Christmas party. I worked on speeches for him, and he read them word for word on the Senate floor. I crammed into the internal subway of the U.S. Capitol with him. Sat behind him during hearings, passing notes back and forth. Joked with him on the couches in his office. My kids like him—he made my older son laugh hysterically the time he knelt on the sidewalk and whispered in his ear. Does this mean I know him?

In elementary school, I imagined that my teachers lived in the school building. Believing in bosses and teachers is part of the joy of throwing myself into work. My spouse and kids also believe— that is why they accommodate my all-nighters and missed family dinners. My work is mission driven, and relationships go beyond a paycheck.

More accusations of groping or forced kisses unfold. I don't *not* believe the women. A far murkier and scarier idea unsettles me. What if everyone is telling the truth, their truth? I wade through an overwhelm of information in my daily life, and social media offers shortcuts to informed opinions. But how do I judge multiple stories of the same events that don't fit together?

I talk and email with other women who worked for the senator. No dark secrets emerge. But everyone is sad. I consider how the accusations rewrite my empowering career story.

When I see my college writing professor, she doesn't talk about poetry. It has been more than a decade, and she is delighted to hear that I still write. But what she wants to talk about is the senator. He made the world safer for women, she says. Even though she doesn't live in his state, losing his service to our country feels personal to her.

It was in this professor's class that my writing developed beyond diary entries and stranded lines of poetry. With her, I learned to shape my notes into art. Her voice still plays in my head. She embraced writing in fragments, letting a single spark or scene come onto a page. She encouraged a process of discovery. Amalgamations of these fragments could capture the larger truths of our experiences. But in her class, she didn't teach us about humor.

Making people laugh can change everything—getting it wrong can change everything too.

Calculations of numbers are absolute, with clear rules, but jokes depend on context. Whatever was the truth in the moment, the delivery, timing, and absurdity have been recalculated. A joke is not a joke is not a joke. A joke is a calculation of the present moment that might not work at another time. *You had to be there.* Jokes fall flat in the retelling. Or worse.

I catalog the subjects of jokes. Bodily functions. Our impending and inevitable deaths. Stating facts, like the unlikelihood of our dreams coming true. The potential of others to harm us. Our potential to harm others. The paradox of systems we rely on hurting us—our families, our schools, our medical care, the police, the government. The existence of brutality, racism, hate, and cruelty. Possible but unlikely risks—the reminder that we can't forget about death by drowning in a foot of water. Jokes delight in our fear, pain, flaws, failure, shame, and differences. What we cannot bear to say—or listen to—in plain language.

For the subject of the joke, it's only funny if they are in on it, safe, a collaborator.

Misconduct has gradients from inappropriate to criminal. Layered on this is the level of distress of the person violated. I do not see a limit to the ripple effects a person might feel when something happens to them. No equation can match a level of misconduct to a level of distress.

I am hopeful that a formal investigation will provide a larger framework through which to understand all this. I want it to delve into the everyday confusion of casual touch: a hand to the shoulder, a hug with too much squeeze, an arm around the waist. Small touches can be platonic, wildly erotic, or invasive. I think an investigation can do what I cannot: hold facts at a distance.

When the senator's close colleagues call for him to resign, the resolution is immediate. But it resolves nothing. The media move on, with miles of fresh headlines to scroll. I am left wanting more than an understanding that everyone has their own truth and their own boundaries and their own sense of humor. That is a reckoning our country is not ready to make.

I watch from within my office as the governor picks my former boss's replacement. I didn't want him to resign. When it is announced that our lieutenant governor will become senator, I call my former coworkers in D.C. to tell them how great she is, all the while gutted to lose our office's guiding star. She will no longer lead from the office across the hall from me. Feeling powerless, I do what is in my control.

I send out an email to all of Senator Franken's former staff, a network that goes far beyond those I have met. It is an offer to join together one last time. My idea riffs off recent public art I've done. I put a book together. Each person gets a page. I ask everyone to send a photo and a handwritten note. The scrawl, scratch, and loops of a pen on the page hold wisps of spirit. As in, I want as much humanity as possible, to make this personal.

The week he officially leaves office, the senator calls me. He says thank you. The book is really something—it's beautiful. It made him cry.

We don't hang up right away, and I don't know what to talk about. I ask about his immediate plans. What does his day look like today? That is the mom in me speaking.

The once a staffer, always a staffer is also present but digging fingernails into my leg so I don't cry. That staffer would like a script. Given the scarcity of time when he was senator, I'd been trained to keep dialogue concise—interactions revolved around specific recommendations, like questions to ask in a confirmation hearing.

"What do you suggest?" he asks.

"To read," I say. My mom compulsively read mystery novels when my Grandma Fanny died.

"Do you have a book you'd suggest?" he asks.

I am sitting on my desk in the State Capitol. I look around for ideas. All I see are policy papers and an open desk drawer full of high heels. I think of a book my husband and I read aloud to each other in our twenties. I'd enjoyed it so much, I had wanted to stay up all night to find out how it ended. *The Golden Compass,* I say. It's the first book in a trilogy with magic, animals, and snowy landscapes.

It doesn't resonate with him. He is a comedian who reads history and biographies. Nonfiction. Overly serious, just as I suspected.

I want to tell him that sometimes the language for our experiences doesn't exist. Sometimes we need entire invented worlds to understand or even see the very real but invisible in our lives—in ourselves. But I don't push it. I suggest *All the Light We Cannot See,* a work of historical fiction.

At the end of the call, he says, "We had so much fun in the office, didn't we?" I feel the same discomfort I felt when I worked for him and people outside the office said, "He must be so funny all the time."

I loved every minute of staffing him and participating in the team he built. I sobbed face down on the carpet when I left. But I would never call working in his office fun. I don't tell him this. It has already been made clear that we can perceive shared experiences very differently.

I lie. "Yes, yes it was so fun," I say.

What I really mean is, we gave it all we had.

Fuck Carpe Diem

"I'm quitting," I say, voice crackling with laryngitis. I've saved my voice all day for this announcement. I am taking control of my life.

My husband, Dan, stands in the doorway, face obscured from above by a furry winter hat and from below by a parka. His and our kids' lives revolve around my being on call twenty-four hours a day and seven days a week—whenever the governor needs me. It's a dream job. Dan looks at me with confusion.

A couple of weeks before, I'd gotten a new supervisor. We have yet to meet in person or establish a relationship. We have not talked about the imminent deadline for flood negotiations with North Dakota and our precariously suspended court proceedings. I've spent the past several months traveling with the governor to meet with a task force in Fargo. I've heard from farmers with crops in fields and families with loved ones in cemeteries that a proposed dam would flood. Governments are watching from across state and national borders. In the next few days, my counterpart in North Dakota and I need to finalize a report on the outcomes of the task force. It is a delicate negotiation of the negotiations.

Earlier in the day I'd sent the new supervisor a quick note. I had laryngitis and would work from home. I need to sit by a humidifier and not talk. Otherwise, I felt fine. My bouts of "Little Mermaid" laryngitis started after my second child was born. While we are expected to make ourselves available on a moment's notice, our previous supervisor trusted us to manage the rest of our time. She believed in us. We were the experts. She is also a mother to young children, and openly acknowledged family responsibilities. We work nights and weekends, but she would schedule meetings after bedtime and during naps.

The new supervisor responded to my note, "Going forward, I would appreciate if you asked, rather than tell me, if you want to work from home. There will be times when it's not possible. I'd also like advance notice."

With that slight tilt in the world, I no longer felt part of a team. *I'd also like advance notice of laryngitis.* I had thought of our office as a community. We supported each other. Leadership had high standards but not tight fists of control. With my new perspective, what looked like striving before now felt like depletion. What I thought of as getting absorbed in the greater purpose of my office now landed as disassociation.

As I wrote my apology email, jagged pieces of glass fell out between the *click click* of my fingers on the keyboard. I held the shards up, and winter light shone through them. They were the precipitate of oversaturation: stress-induced insomnia, sending work emails from playgrounds, sleeping in roadside hotels, not knowing if I'd be home for dinner, and if I was home, not having much left to give.

The Red River flows along the border between Minnesota and North Dakota, and northward to Lake Winnipeg in Canada. In spring northern ice can dam the river, flooding the flat Red River Valley, which was once the bottom of an ancient lake. There are two ways to think about floods: accept that floodplains are wild, or build on them and try to control nature with concrete and steel. In the 1980s Minnesota began buying out homeowners in flood-prone areas. The state respected that water does as it wishes and acted to minimize risks. North Dakota kept building and wants to build more. To allow for expansion, it created plans for a new high-hazard dam and a diversion channel. The two states have opposing ideas about what it means to be in control.

The new supervisor responded to my email: "The fact of the matter is that the way your previous supervisor and others have handled time off is not my style. . . . Things need to change . . . so there will be a shift and an adjustment by all . . . we should discuss connecting with HR RE: FMLA." *Invoke the Family Medical Leave Act for laryngitis?* Since I could not say what I felt, I quoted Amy Poehler to myself, from her book *Yes Please*:

> Treat your career like a bad boyfriend. Here's the thing. Your career won't take care of you. It won't call you back or introduce you to its parents. Your career will openly flirt with other people while you are around. It will forget your birthday and wreck your car. Your career will blow you off if you call it too much. It's never leaving its wife. Your career is fucking other people and everyone knows but you. Your career will never marry you. . . . If your career is a bad boyfriend, it is healthy to remember you can always leave and go sleep with somebody else.

I am no longer sure that putting my career front and center is admirable. I expect a level of care from my office, and perhaps that is naive. Life is short. Last year a peer was diagnosed with stage four cancer while his infant daughter was in the NICU. He died before her second birthday. *Life is short.* If I die now, I will have regrets. I am a writer who hasn't made time for writing. I've published a few short stories and essays, but taking on creative projects is like having an affair. I have to steal, sneak, and secret away time. As a married woman with children and a career, I don't have time for that kind of passion. I shove story ideas and lines of poetry into a locked chamber below my clavicle.

It's been my intent to one day focus on writing. But I haven't known how that choice would show up, and it hasn't. Besides, a jewel locked away is safe from the world. It hasn't been an option for me to not bring in a salary. Dan only recently got a job with health insurance. I see an opening.

Dan, still in winter gear, holds out his arms to me. I step back, anticipating him trying to pull me from the edge. He puts a hand on my shoulder. I want to rant, but I must be judicious with my voice. I say, "We won't have money for day care. Cluck and I will hang out until kindergarten starts in the fall. We'll take public transit and explore the city. I'm going to make us a playlist for the bus!" Dan looks disturbed. I add, "It will be great. And once Cluck starts school, I will write." Fist to eye, I wipe a tear away. I'm not sure if it is one of frustration, sentiment, or relief.

This is not about leaning in or leaning out. If I walk into a room, I am going to sit at the table and speak. This is about doing something far more outrageous. I want to pursue the vocation that persists in calling me, despite my having ignored all its voice-mails. I want to make art.

"I thought you were going to launch into a well-paid job at the end of the governor's term. I thought we'd buy a bigger house." Dan says this without emotion, like he's assessing tomatoes at the store. He is not one to react strongly in the moment.

I think, *I am good at whatever I do. Really good. As long as it's for someone else.* The thought is like a stone dropping through a pond and landing with certainty on the silty bottom.

Dan pulls me close. *Why not do what I want?* I feel the weight of that stone in my hand. "Fuck carpe diem," I whisper shout.

"I don't get it," Dan pulls back. His eyebrows furrow.

"I don't care about houses."

"There isn't room for the kids to play with friends when they come over," he says.

"I don't want to be infinitely energetic and productive. I don't care about a fancy career and a well-paid job. I don't care about power. *Fuck carpe diem.*"

"Ummmm, I still don't get it," he says.

"I just want to sleep well at night. I want to be present when I'm with the kids. I want to write," I say.

"That sounds like carpe diem to me."

"No. Carpe diem is go big or go home, work hard and play hard, burn the candle at both ends, worry the kids are growing up so fast and I have to catch every moment." The more I try to shout, the more the words evaporate in my throat, but I can't stop. "Carpe diem is having career success and babies and stay-ing fit and going out with friends and visiting family all over the world, and, and—it is too much!"

"Anna, will you please just think about it longer?" he asks.

"*Fuck carpe diem* is about appreciating the negative space.

Music is the silence between the notes," I say. Dan and I met as geology majors, we are scientists who study time. Nothing is worth more than our time.

"Think about it?" he says.

"Hmmfpf," I snort. My friend Natalie told me she makes decisions by changing her mind. Fully decide, try it on, and then walk away. Toggle back and forth. I look up at him and nod.

With the Red River, concrete-and-steel defenses provide protection against flooding, but the idea of total control is an illusion. Levees break. Floodwalls collapse. People get stranded. Dams hold water back from some areas while flooding others. Minnesota's approach, to get out of the way of the river, acknowledges our limited power. It is a kind of flood protection through relationship. By giving up some control and accepting that the river needs space, we reduce the risks communities face.

The governor gives us each a proclamation for our birthday. My new supervisor reads mine aloud at a staff meeting.

> Whereas Dr. Anna Henderson has brought an immense range of technical, practical, and public policy expertise to the Governor's Office, and . . .
>
> Whereas Anna's unofficial title, *Water Czar*, strikes fear in the hearts of nitrates everywhere; and . . .
>
> Now Therefore, I, Mark Dayton, Governor of Minnesota, do hereby proclaim Saturday, March 3, 2018, as Anna Henderson Day in the State of Minnesota.

When she finishes reading, she laughs, "Anna, you have a PhD?" I nod.

"That is the most hilarious thing I've heard," my supervisor says.

The new supervisor's email pings on our computers and then there are the soft thuds and clicks as office doors close. She outlines her policies: be at your desk from nine to five. If you have off-site meetings with agencies or stakeholders, if you are sick and want to stay home, if your kids have an emergency and you need to deal with it, if you worked fifteen hours the day before and want to come in late—it is her decision to make. We creep from office to office like children sneaking out of bed. I'm relieved. I had felt personally attacked before, but now I understand this is not about me.

"I'm going to quit," I hiss to a colleague.

"Before or during the legislative session?"

"Shit," I sigh. Session starts in only a few weeks. "It would have to be before to make it worth it." I picture myself sitting with my five-year-old on a city bus listening to music on our headphones.

Before, most of my conversations with colleagues were about a particular legislator or stakeholder. Now we talk about how the supervisor was bad-mouthing our team at a happy hour. *Can you believe she shit-talked him for taking paternity leave?* We pause to ask each other, *Are you okay?* We linger. We talk about our families, a decision to give up alcohol for Lent, a favorite book, a dream of summer tulips. Out of bitterness comes the sweetness of getting to know each other more as people.

The governor goes for my idea to require testing of well water when a home is sold to ensure that the new residents are aware of any contamination. It is something I can do to protect public health. I have been given a charge: I must stay.

My colleague is rebuked for leaving early in a snowstorm . . . *You didn't ask my permission* . . . I decide to quit.

I go back and forth. See how it feels.

I no longer come in at seven or eight in the morning. I pull back. The need to seek permission, to explain each meeting, to justify doing my work—the constraints under the new supervisor conflict with my ability to take care of my responsibilities, myself, and my children.

In 2016 Minnesota denied North Dakota's permit to build a high-hazard dam—the "high-hazard" classification meaning that dam failure or misoperation could result in loss of human life and property. North Dakota and the U.S. Army Corps of Engineers claimed that the federal government could override state law. They went ahead and signed contracts to start construction. Minnesota sued, a court stayed construction, and North Dakota has been paying contractors to do nothing ever since.

When Minnesota won a court case about the dam in 2017, the governor used the win as leverage to create a transparent process. To begin to make a compromise he flew to talk with residents along both sides of the Red River. He quoted the late Minnesota senator Paul Wellstone, "We all do better when we all do better." I hadn't worked on the issue before, but I flew with him to Fargo every week in a soda can of an airplane. Motion sickness made the world spin and my skin tinge green for hours after each flight. Even though we had won, we still had to work things out. The river would flood in the future, and people needed protection.

"Do you have a moment to speak privately?" I ask our chief of staff.

The chief, my former supervisor, looks at me with concern. My voice and body tremble. This is someone I don't just like but admire. I don't want to be a troublemaker. Also, if I say it out loud, it will all be true: I no longer feel safe here. "I want to talk to you about the new office policies." I have cut out as few words as possible from my internal tirade and fit them together without glue or nails or staples.

While we deal daily with confrontations and pushing boundaries, these are usually with forces outside our office. Fascinated and horrified, I watch red blotches bloom on the chief's face and neck. She lets me know she hears me, makes no promises, gives nothing away. It is the way we respond to lobbyists.

Effective supervisors, leaders, know that you are the expert in your job. They serve as guardrails. You walk into their offices, download your progress, concerns, and plans. They say forget this, or sounds great and make sure you do that. I walk into this woman's office, lay the weight of my burden on her, and walk out free to do my work. This changes nothing and everything.

No policies change. But asserting my humanity lets me set aside frustration. I consider my future not as a reaction, but in the broader context of my life.

In Fargo, I sat with the task force: people vehemently opposed to flood control, people in full support, people from cities, and people from rural communities. Half were from North Dakota, half from Minnesota. The two governors chaired the meetings in front of cameras, with open seating for the public. The only thing task force members had in common were stories of *flood fighting*. Everyone had their lines drawn in the sand. They were dug in.

Between the formal meetings, I sat with engineers and traced maps with my finger. I touched the homes, cemeteries, and farms that the plan would protect, and the ones it would flood. I called and answered calls from members of communities in both states. Sometimes they raged for a long time. I listened, and I waited for the clearing.

The trick to moving negotiations forward was allowing people's emotions to take up space. Only after this could we talk about letting the river rise higher, adding structures and expenses that would reduce how much land the dam flooded. The meetings did not create unanimous consensus, but everyone gave a little. Extremes fell away. As in all negotiations, the resulting compromise satisfied no one, but it offered a way to move forward.

Forgiveness is not about liking my supervisor or feeling good about how she treats me. It's a trick of perspective. It is seeing the smallness and vulnerability in both her and me. It is seeing the true limit of her powers so that she can no longer walk through walls and peer over my shoulder.

Forgiveness is also loving her bright blue and orange eyeshadows. It is delighting in how she Photoshops her dog into exotic vacation photos.

I can't say if her rules give her a sense of control, or if they make her feel powerful or happy. But I do know that each time I break them, it gives me pleasure. Like with a river, no matter how much force is used to control it, it will only ever be what it is. Through small acts of subversion, I slip back into a kind of normal. I grant myself permission to do what needs to get done.

My initial shock turned to anger, and anger to excitement. The problem is not just my supervisor. The problem is that I have been sucking my marrow dry. *Things need to change.* My life is not in balance. Or the effort to keep balance has been too violent. It is physically painful to rip myself away from the lull of children eating breakfast and imagining dinosaurs in the Mariana Trench. A thin thread connects my body to them, and when I walk out the door that thread stretches beyond the breaking point. At the end of the workday, I tear myself from an urgent tangle of memos, strategies, and negotiations. Like an alcoholic gone dry, I'm left slightly shaky. I am in a cycle of burnout and recovery, with the recovery never lasting long enough.

The demands of my career have armed me with a shield I use to avoid looking into the abyss. But maybe there is something to peering into the dark depths. To not being busy all the time.

I decide to jump, but to *plan* my jump. Instead of taking an escape hatch, I focus on growth. I make a commitment to stay through the end of the administration, eleven more months. That gives me time to prepare for the next phase of life. I hold myself back a little more each day, test out saving more of myself for myself. On Sunday afternoons I go to the library to write. I turn off my phone. For a brief period, I am available only to myself. Life is an accumulation of choices. And I've made up my mind.

Packing for an Overnight at the State Capitol

No one likes conflict, but with the smack of a fist I am a million particles of brilliant light. However, tonight, I'm taking the punches. The letter from the House and Senate chairs of the Agriculture Committees is a direct threat to the governor's water agenda, a blunt whack to the nose. I haven't been home for dinner in days, and I can't remember what it feels like to help my boys into their pajamas. I'm tired and mad and, for a moment, frozen in place. It's Friday, well past the mid-May sunset. My life has been reduced to a countdown to the end of the 2018 legislative session Sunday at midnight. Firm in my commitment to leave politics and write, I am seeing the last legislative session of the administration through.

I jump up and look into the hallway of quarter-sawn oak doors. Realizing I'm barefoot, I grab heels from my bottom desk drawer.

The door cracks open: Tenzin's long black hair and heart-shaped face. She pulls me in.

"When we look back, won't it be obvious this was another *Flint*?" I say.

"We shouldn't negotiate," she winds her arms into the thin wool of her white shawl.

I smile, relieved that at least she and I won't be battling each other.

Tenzin grew up as a Tibetan refugee in India, where her well often ran dry in summer. I don't have to convince her that safe drinking water is a choice we make over and over.

When she immigrated during high school, I was finishing college. I admire her political instinct, and though she is younger, she mentors me. In our jobs advising the governor, our peers are our best mentors. We can't trust the motives of anyone else.

"Their constituents don't believe drinking fertilizer can kill babies?" I ask. While the governor's signature buffer law protects lakes and rivers, it doesn't prevent contamination of the groundwater we drink.

"It isn't about that," Tenzin shakes her head. I notice the big circles under her eyes. I wonder if I look as worn out as she does. Maybe worse.

"Studies show links to cancer in adults, too," I say. In 1989, the Minnesota Legislature recognized that chemicals used in farming pollute drinking water. It is 2018, and we have yet to fully implement the law passed in the eighties to address contamination from fertilizer chemicals. Our administration has spent years convening stakeholders to inform a rule regulating fertilizer application. Last year we had seventeen public meetings across the state. The rule is ready.

"They need a win," Tenzin says, and hands me a bowl of candy bars. I taste only sugar and salt. I picture the House and Senate Agriculture Committee chairs dragging dead carcasses down their main streets.

The letter from the chairs offers the governor an ultimatum: if he doesn't sign their bill gutting part of the buffer law, the legislative committees will kill his new rule to protect rural drinking water. Their threat relies on an arcane, never-before-used law. With it, the committees can stall the rule—if their party wins the fall election, the rule will die. If we don't take their offer, we risk losing the chance to regulate fertilizer at all.

Our phones vibrate and we look into our palms. The governor. He wants a draft response.

"I'll take the first shot," I say and run out. Agriculture is Tenzin's portfolio, but the miasma of the buffer law is mine.

I dash downstairs into a deserted hall of the Capitol to sit on a bench. A gargoyle on the base of a lamp gives me an I-just-bit-a-lemon face. To get close enough to knock the wind out of the committee chairs, I write in longhand.

"I am shocked and—" I cross out the words. I used that phrase only a month ago.

"I'm appalled that you are holding a hearing so that you can deny rural Minnesotans their rights to clean and safe water." The words flow from the ether of the building.

I run upstairs. Edit as I type. Print. Read aloud. Edit. Repeat.

Tenzin sits at my computer adding her own words. We pass the keyboard back and forth, no laughter, no swears. We're channeling something deeper. Together, our fists join for ultimate impact.

Things to pack from the kitchen:
cut fruit, cut veggies, cheddar cheese.
Note: Slicing these gives a sense of control.

I'm tired. I hurt like I worked a double shift waitressing. I woke at 5:00 a.m. churning the hundreds of faces I passed in the Rotunda yesterday, the tens of people I met with, the endless emails I pounded out responses to. Now, I take in the sounds of my young boys playing a game of sea creatures. I take in the smell of coffee and toast. I need to go back to the Capitol. It is Saturday morning, and I'll likely be gone until Monday. I have to pack.

"I found the Lego!" my younger son runs into the room. He looks at me, his face falls. He's been warned not to wake me. I motion with my hand and pull him close, breathe his warmth. As far as work-life balance goes, right now it's all work. He rubs his cheek against mine like we're bolts of silk.

My older son comes in. "Mama?" he says, but the question in his voice dies off. I sit on the edge of the bed. Both boys hold on to me. I lean into their wild curls sticking up in all directions. I'm tethered. There are things I pack without realizing it. Tenderness

I will discover later and marvel at, but that kind of unpacking won't happen until I leave my job.

"When did you get home?" my older boy asks.

"Late," I shrug.

Things to pack from the bedroom:
sweatpants to go under suit dress,
toothbrush and toothpaste.

I transform before I get out of my car. Last-minute item to pack: my smile. Really, most facial expressions. Along with these I pack my desire for a family bike ride, a video call with my niece, coffee with a friend, curling up with a book.

I'm usually quick to smile and quick to cry. Good news or bad, it hits me like vinegar on baking soda. But at the Capitol I keep to a narrow range of emotions. The rest I put in the trunk of my car.

Things to pack from the camping section
of the basement:
sleeping pad and pillow.

An invisible umbilical cord reels me down the hill from the parking lot. The sun is out, and the white marble of the building hurts my eyes. The circle of eagles around the Capitol dome looks ready to break free.

I start up the wide stairs to the main doors and realize it will be like an airport terminal in a snowstorm. The waiting lobbyists and activists will want updates. At the Capitol, information is currency. Its distribution forms and breaks bonds. I weigh the balance of engaging versus avoiding. Engaging could avert misunderstandings and keep lines of communication open.

The State Capitol has the fishbowl effect of high school: too many people smashed together. As in high school, clothing is a coded language that signals who you are. Fresh and in style—corporate or philanthropic. Outdated suits that smell of sweat—lobbyist or legislator. People in jeans, fleece vests, or leather jackets—advocates for everything from stopping mines to preventing helmet laws. Older people in coordinated outfits—tourists.

One weapon of the majority is to set the schedule and not share it. This morning both House and Senate members have a roll call. But after? Bills could come to the floor for votes. Or leadership could go into a room to slam together their giant spitball behind closed doors. This megabill, the omnibus-omnibus, will combine all program funding and cuts with all policy changes in all areas. Immigration policy with chronic wasting disease in deer with wastewater treatment. A shit show. If you know the schedule, you know when you can nap and eat. Exhaustion and hunger erode resolve, make what was not possible before, possible. Members of the minority party are forced into battle with their own bodies. At some point closing your eyes is all that matters.

I veer away from the main steps and go through the unadorned doors of the ground floor. No vines or branches decorate the bare limestone walls. My right eyelid thrashes, refusing to let me ignore exhaustion.

Will we negotiate? If they can pull at the governor's heartstrings, if they are respectful, if they sit face-to-face with him—I don't want to think about it. I've seen him fold.

I don't realize I am holding my breath until I enter my office. I notice honey stains on the shoulder of my dress. I'm ambivalent about washing off these sticky paw prints of my boys.

Hearing excited chatter, I step into the hall. My colleagues are watching Prince Harry and Meghan Markle's royal wedding.

"Serena Williams!"

"Elton John!"

"Oprah!"

We wonder at the whimsical hats and the Gothic buildings of Windsor Castle. "Scones and clotted cream outside the communication office!" a colleague shouts. A TV on the other side of the room shows House members assembling, but I turn away to watch the thousands of waving British flags. A surge from across the ocean. It's a tailgating party.

Things to pack from cabinet members: the "honey badger don't care" attitude.

Tenzin and I, we have the perspective of being outside the state agencies and seeing them with a bird's-eye view. The group of environmental agency leaders seated at Tenzin's table, they have the institutional knowledge. Risk averse, they come as a united front against the governor. They want to take the Ag Committee chair's deal and let part of the buffer law be cut away.

"If we offer them—" one starts. He is lean, strong, and feral. His hair speckled gray and cheeks hollowed with the first tinge of old age. When he agrees with us, he is the best. If he doesn't agree, he smiles and then does what he wants. He and I have been in a game of cat and mouse all session.

"No," Tenzin cuts him off.

"We need to offer something," another says. He's well-groomed, but not flashy. Always polite. Refined. He speaks in a quiet voice, only his urgency to jump in betrays his anxiety.

I stand up, pull on my suit jacket, say nothing, and maintain eye contact. *To honey badger* is a verb. All I need to do is be still. No scowls. More important, no smiles. If my poker face is pulled correctly taut, threats and laughter will ding like hail on a metal bucket.

"They're blackmailing us," I finally say. The music from the royal wedding floats through the half-open door. For a moment, I entertain a fantasy of stealing the box of scones and jar of clotted cream. I'd hide in a closet and eat them one by one. I can't imagine anyplace I'd rather be.

"They need a win!" an agency lawyer shouts. Her smile is mismatched to her exasperation.

"This isn't a game. It's public health," I snap. I remind myself: arguing back shows weakness.

Those motherfuckers, I think, *their blackmail will break us apart.* While the fight is with the other political party, the only people I am *fighting* with are on my own team.

"Anna, they have to have something to show. They can't go back to their districts without another shot at the environment," the refined one says in his calm voice.

The honey badger instinct is natural with opponents. To do it with my own inner circle sends prickles of heat across my body. *If I am inert steel, he will get nervous.* We are animals made to mirror each other. I let his words sour in the air. Politics is a war of endurance.

The honey badger, *Mellivora capensis* in Latin, also known as the ratel, a name used for a South African armored military vehicle designed for rapid offense, has a literal thick skin. It can withstand beestings, porcupine quills, machete blows, and animal bites. In a viral Internet meme, after a honey badger is bitten by a poisonous snake, it passes out, wakes up, and goes back to eating. *That is who I need to be.*

The governor's cabinet hadn't initially wanted him to unleash a new environmental regulation. They knew it would be a battle, and they would be on the front lines. They've traveled the state and fought for it, but I still don't trust their impulses. With the administration ending, we are all on edge. All about to go on the job market.

Tenzin jumps in, "The governor was crystal clear, we're not negotiating. None of you are to negotiate. He sent his response, and now we wait."

"Of course. We all get it," the feral one starts up. "We aren't negotiating. But we need to be ready. Really, this part of the law—"

"It's not worth keeping if we have to lose something else," the lawyer jumps in. She giggles in apology. As soon as she quiets, her face pinches back to a pained look.

"How is the commissioner?" I ask. The Department of Agriculture commissioner's daughter passed away only days ago. Earlier in the week, we'd stood in back of a crowded room, crying for someone we'd never met. Someone who meant something to us because of how we feel about the commissioner.

No one plasters a smile on their face now. "It's hard," the lawyer says.

Things to pack from the children:
green ninja warrior figure
so they will stop fighting over it.

Back in my office, the shouts, songs, and clatter of footsteps come through the walls of the Rotunda. The sound carries, but the marble and oak distort the words. I imagine this is what it would sound like to listen to someone else's dreams. I stand at my desk. I have no hunger, no need for sleep. I have no memory of bathing wiggly children, wrapping them in towels, or drying their curls. All that is distant. Here I'm part of a different and larger organism. I feel something in my dress pocket and pull out a plastic ninja. I place it on my computer to watch over me.

I'm alert to the *ping ping* of texts and emails dashing through air currents. The House debates a bill to dismantle government programs on my officemate's TV. The Senate votes to overrule a judge with legislation on my TV. My officemate's phone rings, then stops, my phone rings. I check the number, look across the room to my officemate, and we pass a knowing look between us. Everyone wants updates. We share a smile. I shrug. Neither of us picks up the phone.

Without warning I flip the telescope the other way. The room gets very small, words on the page are ants. The State Capitol and everything in it is tiny. I think of the blue sky outside, the way jokes with a four-year-old are silly, not cynical.

I pour a rainbow of Skittles into my hand, take a breath, and slowly force the telescope around so the photographs of my boys blur and the room snaps into focus.

I'm at the printer when the feral one walks by. "Oh, hey Anna."

His smile verges on flirtatious. A habit from lobbying, and nothing to do with me.

"So, what have you been up to?" I ask.

"Well, we were talking to the Agriculture Committee chairman. Really, what we have to give up from the buffer law isn't a big deal."

"You were talking to the chairman?"

"We were just talking. It's good to keep the lines open."

With that impact, a wind rushes through my head.

Things to pack away unconsciously: smiling in professional settings unless it is a strategic tool, public tears, the existence of children in professional conversation, any suggestion that it is not normal to work without knowing when you will go home.

I trail behind my supervisor. She wears a polka-dot sweater and holds a polka-dot water bottle in one hand. The cuteness of dots contrasts the way she stops in the doorway. Legs wide, hands on hips, she arches an eyebrow.

"Did you get clear instructions from the governor to not negotiate?" She looks directly at each of the agency leaders. "So, tell me, why are you talking to committee chairs?"

I look at Tenzin but she shows nothing.

"We weren't negotiating. We were just talking. It's good to keep open the lines—" the feral one talks fast.

"Well whatever you call it, let me be clear. If the governor wants you to do something, he'll tell you." She walks out with an audible sigh of disgust.

"Anna?" the refined one bites at my name. "What the *fuck* are you trying? What the *fuck* is going on? This doesn't *fucking* make any sense." I match Tenzin's blank face though I feel a fire in my body. Fists clench.

I think of the time this man and I drove to meet with a business executive in a western suburb. It was an icy winter. Halfway up the steep driveway at the executive's mansion, our car stopped. As if we could will it with our bodies, we both groaned, but the car slid back into the street. We tried again, slid back. Finally, we gave in. In it together, we laughed in the car, parked on the street with no other parked cars and no sidewalks, and walked up to the door, uncertain of what to expect.

Now, the tension comes off him like the crackling of an electric fence. Something in me shifts. I've always told myself that I'm just the governor's messenger. But this, this moment, it's like when my brother jumped onto the seesaw and sent me flying into the air with the smack of the wood plank on my bottom. This man, he's never been my friend. None of them have.

I take a breath and say to myself, "Honey badger don't care."

Because my heart is thudding and inside my head roars, I walk out. I stand against the cold marble bathroom wall until the anger settles. Until I can haul it up the hill to the trunk of my car where the rest of me is packed away.

Where the Night Stars Are Orchids

Sine Die: The Constitutionally Dictated End of the Legislative Session at Midnight

The thought strikes after I stand on top of the dome of the State Capitol. It is the last night of the legislative session, and for a few moments the stacks of paper and reek of sweat are two hundred feet below. The air outside so fresh I could eat it by the handful. Car lights and streetlamps glimmer like jewels. We've come up through the dust of rock and brick between the interior dome one looks up at from the Rotunda and the exterior dome one sees from the street. It's an in-between place in the skeleton of the building. Below us every lobbyist, nonprofit director, and activist waits for Minnesota's 134 representatives and 67 senators to finalize the budget and vote.

Life flows back into me from the stars, the sky, and the slapstick feeling of sleep deprivation. I walk away from the laughing group to circumnavigate the cupola, pause at a ladder to the very tippy top. It feels possible, certain, the higher the better. But I am unsure of my footing in heels—the least favorite part of my work costume. And the tug, like an umbilical cord, even out here is too strong. In the land of ten thousand lakes, my position as water adviser to the governor is, in and of itself, an expression of our public health and environmental crisis—and the level of contention over doing anything about it.

Back in the bustling hall of lobbyists and legislative aides, our group pulls into a huddle. "Watch for the final bill to post, and—" my boss says, but I don't hear the rest because I'm looking up.

"Hey," I interrupt, "do you know what plant that is?" Everyone looks up at the leaves overlapping like fish scales around a light fixture. The spark had been there for months, maybe even since the beginning, but I was too busy to notice. It might be less that the spark takes off and more that there is finally a clearing inside myself for the fire. When you hold your breath, there is no oxygen to burn.

"Watch for the bill," my boss says, again. We are dismissed. The others scatter, but I stay where I am. I spin around, take in the pungent fruit and grain and flowers painted on the walls of the Capitol. The trained scientist inside me is back on a mountain expedition and needs to know what story these plants tell.

In the fleet of agencies that make up the administration, the governor's office, where I work, is the flagship. We are docked in the west wing of the Capitol. The hallways are narrow galleys, each of the three floors its own deck with its own function: policy on top, then legal, then communications. Government departments are spread across the city, each run by a commissioner appointed by the governor to fill his cabinet. Cabinet members make their own rules on their own ships.

For the grand finale of the legislative session, all the captains board our ship. Together we work out how to respond to texts, emails, calls, letters, spreadsheets, and media reports—these are the mediums for threats, reconciliation, and negotiation with the legislature. We spend a lot of time waiting for the next thing to happen. Sometimes we share keyboards and red pens. Other times I send the captains out to confer with my own shipmates. We all work for the governor, and yet our goals don't always align.

I enter the ship on the lower deck. A captain is wiping tears from red eyes. I look away. Passing a darkened office, I hear whispering voices, familiar ones. Good friends or lovers? They are both married. My gut says lovers. It's the trading back and forth of words as they explain survey results, the quiver between them, the way they form a wall of defense. I have a question for them, but I don't go in.

A text pings, the final bill has posted. It is twenty minutes to midnight. I make my way through a crowded hall of crunching M&Ms, soda cans snapping open, and trails of lettuce from walking tacos. I pick a bright office where people sit on the desk, lean against the wall, and scrunch together on window ledges. The House debates on a TV on one side, the Senate on the other. The volume is up on both.

Minutes tick by. Legislators swim upstream. They speak with force, absolutes, and certainties. They point out that no one could have reviewed the hundreds of pages of the bill in the last few minutes. They demand transparency as representatives of the people, but the clock is out of time. I am sick to my stomach. I want to run far and fast.

They vote. The budget passes. Midnight chimes. It is *sine die*.

The governor has the last word: after the shouting of hundreds of people at rallies, after thousands of letters and calls, hourly media reports, editorials, legislative committee meetings, floor votes, fear, hope, and games of chicken—the hundred pounds of cheese and the thousand cookies consumed in our office over the past ninety-six hours—the governor stamps a bright-red VETO on the omnibus-omnibus comprehensive bill. Session is over. All the work of the past months has led to nothing.

In Search of a Fuel Other Than Adrenaline

It's 3:00 a.m. when I get home. A pitter-patter and my eight-year-old stands in the kitchen doorway in Spider-Man pajamas.

"What are you doing?" he rubs his eyes.

I'm too tired to take his hand and walk him back to bed so I open my arms. He climbs onto my lap, and I listen to the cadence of his sweet voice.

The switch in pace and emotional range from work to home is jarring. I hug my son, smell his hair, study the peeling skin on his upper lip. I try to respond enough so he will keep talking.

I dream of spreadsheets and wake rehashing conversations—my mind a broken record. Breakfast with my family is like speaking a language I'm not practiced in. I know I will help the kids button shirts and get shoes on, but right now I can't remember how. I whisper stage directions to myself. I still can't believe my luck, *I'm a mother and these are my sons.*

A swaying sensation lasts in my body for days. I don't have my land legs. I'm hollowed out, desiccated by salty chips and Sour Patch Kids I've used like breadcrumbs to lead me out of the night and into each brand-new day.

The senators and representatives return to their farms and jobs, the lobbyists to their offices. But the Capitol continues to be a place to gather. An activist tells me that he is always studying its production possibility. In my phone notes app I jot down:

> Indigenous dancers with jingling bells on their ankles.
> A group of monks like bright-orange flames.
> Two girls run up the steps, drop their towels,
> and take photos in string bikinis.

An audience at the Capitol can be a crowd so large it spills onto the front lawn and stops traffic, but most often the only witness is a phone camera and the building itself.

Now that session is over, everything is different. We know, the media know, everyone knows, this is the beginning of the end. The governor's term ends in seven months, he will retire, and our paracosm will fold back into the marble floors and walls of the Capitol. Soon we will be watching from the outside as the mast of a new governor's office bobs on the horizon. From whichever party, it's doubtful the next administration will continue to focus on water.

Water *is* life, and *life* is just too political: factory farms, pipelines, mines, industry, power generation, winter de-icing, farm fertilizer, lead pipes in schools. The list goes on. I have seven months to make sure laws are implemented. Seven months to tie knots that cannot be undone.

I slip off the ship at lunch. I need an anchor. I need fuel that makes me feel whole instead of spent. I don't walk bent over my phone. I look up at the Capitol. The building's motifs of waves, spirals, and zigzags represent natural forces of wind and water. These, along with fruits, flowers, and leaves, cover the walls and ceilings. Designs call attention to windows, doors, and archways. I take a sketchpad and a pencil out of my purse. My contraband. Beauty is not part of my role as a policy adviser.

The Capitol's architect, Cass Gilbert, wrote to a friend in 1921, "Sketch everything in sight . . . no matter how badly you draw, continue . . . keep the pencil active and the mind will keep pace." I stand in an alcove. I don't know how to start. Vines curl into each other, scrolls written in cursive by a hand that never lifted off the page. Pencil to page, like jumping off a cliff.

The building accentuates sound. A closing door becomes a rainstorm, feet on the stairs a flash flood, children racing a mob. Sounds slam against the marble walls until they shatter into a wordless roar.

The Capitol's painter, Elmer Garnsey, wrote of the building, "Mysterious shadows seem to float out of the arched recesses, advancing and receding with weird effect." I bring in binoculars to look more closely. I see the obvious: the same deeply lobed leaf wrapping the bases of cornucopias, on the tops of columns, and painted on the walls.

Before politics, as a scientist, I read the marshes along the Atlantic, the mountains in the Rockies, the walls of an impact crater in Ghana like the pages of a book—pages made of mud, plants, and lakes. I walked. I drew what I saw into my field notes.

I want nature to have been the raw material that inspired the climbing ivy, corn, oak, and pine on these walls. Lewis F. Day wrote at the end of the nineteenth century, when Cass Gilbert was designing the Capitol, that designers were "reflecting reflections of reflections, as though we dared not face the naked light of nature." I have become a reflection of a reflection of a scientist. My relationship to nature is a memory of a memory.

I use a scientific field guide to find a name for the deeply lobed leaves: *Acanthus*. A plant native to southern Europe and northern Africa—a plant that could not survive a Minnesota winter. The State Capitol combines the Minnesota landscape with invasive species from the Italian Renaissance. Or the other way around.

With a name, I see acanthus everywhere. An echo of an ancient Greek story about acanthus leaves growing on a grave. A symbol of immortality. Today we wrap columns and lampposts in foliage, a skin of infinity, a desire for no ending, forever and ever.

Mythological Animals and Other Reminders of Accountability

I let the air hang heavy between us. The first answer in government is always *no*. Even on the governor's flagship, *especially on the flagship*. While the cabinet member I face also serves the governor, he is on the defensive. The space between the governor's flagship and the governor's fleet breeds paranoia. That is the space I inhabit.

No today is in the form of, "Why are you trying to destroy all of our programs?" I had asked him to work with a council of citizens. He has been at war with this council for the past eight years. It is a mess with a hundred million dollars a year for water protection at stake. *It's not personal,* I remind myself and smile into the five feet of silence between us.

Today, dominance between this cabinet member and me will be determined by who is most grounded. It is a staring contest. Seconds tick by. I ignore all body alarms: heat in my neck, wave of dizziness, knot in my stomach, twitch in my left eyebrow. My body shouts, *We are pack animals.* But, in this competition, winning means going it alone.

I've been doing this longer than you've been alive, another cabinet member had yelled over the phone the month before when I called about a deadline.

The cabinet holds power publicly. Commissioners command their own staffs and are quoted in the media. My power is a different kind of currency. I talk directly to the governor, not the press. I work through the cabinet members' spreadsheets during budget crises, manage their lawsuit negotiations, and monitor their press emergencies. When one cabinet member refused to write a public apology earlier in session, I had him sign his name to the one I wrote. I also wrote the governor's rebuke. Legislators quoted both letters in hearings. It was like the old cartoon of Bugs Bunny playing tennis with himself.

After the meeting, I walk through the ground floor and touch the cool yellow-gray limestone. The walls and floor hold multiple stories of sea creatures and ocean tides. When the limestone seafloor was pushed up into mountains, it turned into the green-and-gray marble of the upper floors in the Capitol. I hear older lobbyists talk about bone spurs from years of running on the building's hard surfaces.

Up the stairs to the first floor, I make my way through an arbor of grapes. This is the public entrance to the building. It is a garden with cornucopias and lady's slippers, the state flower. Buff, dusty rose, blue, and green. It's a Dionysian invitation to the public. A welcome from the nymphs and fauns engraved in the lamps.

I stare into the eyes of the griffin guarding the front doors. I nod to the winged lion. The sirens, sphinx, and other mythological animals are reminders to staff, lobbyists, elected officials, judges, cabinet members, and reporters. It is our responsibility to endure, to fight, to speak, to not be swayed by compliments or threats or the desire to be liked.

A Woman as a Parable

In the Rotunda, my gaze becomes the hands on a clock. East, south, west, north: in each archway cloth drapes a woman with naked cherubs and seasonal crops. Her lack of expression could have been copied from a stone statue. In the dome, astrological symbols flanked by cherubs depict the months. A calendar with time repeating as infinitely as the twists of the Greek keys intertwined with vines on the walls. The hand of time passing by. My time, all of our time, is running out.

While my focus is on the plants, the books about the Capitol and the tours of the building center on human figures. The people on the walls are all white. The women are poker-faced, expressionless, even the one with a fox head. The woman-with-an-animal-head represents sin that the "American Genius" must withstand. The genius is a white male underwear model. I don't see myself in these women. They aren't protagonists. They aren't even people. They are sin, industry, virtue, wisdom, fear, and the past, present, and future.

The people with whom I meet are disproportionately white men: legislators, heads of nonprofits, the captains of the executive branch. I see *their* narrative reflected in the paintings, but I don't see myself or a reflection of what Minnesotans look like.

Echoes of Antiquity

I have three emails open and a memo due in two hours. I stack binders on top of piles of paper to make room for sticky notes reminding me of urgent tasks. My phone rings. The voice tells me about a shortage of lab capacity across the country. Meaning, the testing rate for per- and polyfluoroalkyl substances, PFAS, in drinking water will be slower than planned. Meaning, there are people who won't know if it's safe to drink water from their taps.

The day the legislative session started in February, the state settled a lawsuit for $850 million for damages to natural resources and drinking water. We are ground zero for 3M—Minnesota Mining and Manufacturing—a Fortune 500 company. It feels personal, but when I talk to colleagues in other states, I realize there are many ground zeros.

Two days before the court trial was set to begin, I backed into the forest on my cross-country skis and called in to a conference with the governor. I had only five minutes' warning of the meeting. The call-in number too long to remember, I wrote it in the snow with my ski pole. All sides were sure a settlement couldn't be reached. By 9:00 p.m. the next day, the idea of a settlement was like smoke rings that vanish as soon as they appear. At 4:00 a.m. we had an agreement.

I pull up an aerial photo showing the areas of contamination. I click on the purple dots where wells have tested unsafe. The safe limit ratchets down lower and lower as health experts study impacts on breast milk and developing fetuses. I can't picture the faces these dots represent, but each one is a stone crushing my shoulders.

I push back across the phone line. This might not be a problem I can solve with force, but I don't know what else to do.

Normally, I would drop a paragraph with an update into the governor's daily memo, and then turn to the next item on my list. Today, I pause. Instead of diving back in, I grab my purse and walk off the ship.

On the second floor, I peer into the Supreme Court, the House, the Senate, and the small corner store that sells ice cream and popcorn during session. It's here that legislators mingle before votes, and lobbyists prowl with handshakes and smiles. When the building was first painted, the walls were described as Pompeiian red, but to me they are Marilyn Monroe's lipstick. The dominant species is metallic-gold laurel. The literal landscape of laurel is Mediterranean: hot, dry summers and mild winters. Not a midwestern plant. But as an icon of antiquity, laurel connects past to present, affirming the state's identity.

I wander among the columns of the east and west wings like they are trees in a forest. Looking for information about the flora and fauna in the building, I go to the library. In Sally B. Woodbridge's *Details: The Architect's Art,* I read how Greek temples may have started as altars set in groves of trees, and these would have been decorated with garlands of flowers, fruits, and animal sacrifices.

From Woodbridge, I find art historian George Hersey, who hypothesized that the proportions of columns relate to the human form: the fluted pattern of the columns like the folds of cloth, the top of a column the head, or *capital,* below this the throat, or *trachelium,* and crowning it all a leafy band that could be coiled braids. The design of such columns, ubiquitous in the Capitol and many buildings, may derive from images of human sacrifice victims lashed to temple supports.

I want to think that we have created a less cruel society than the Romans. We don't sacrifice gladiators for entertainment. But we do test the safety of more than eighty thousand chemicals used in industrial processes, toys, household products, and cosmetics as an ongoing experiment on the public. Only after chemicals are connected to cancer, illness, and family tragedy do we begin to discuss their safety. The human sacrifices are unintended but predetermined.

Tarot Cards

Settled against a column at the entrance to the empty Senate chamber, I look around for a plant to pull out of the swirling weave. Purse with drawing supplies over my shoulder, I breathe and look around. Deciding on ivy, simple lines I'm confident I can render, I follow the tip of the pencil across the open space of my notebook. Vines fill the page.

The tarot cards for ivy are many. It is associated with death, as vines growing on trees can weaken and kill the host plant; with undying love, as the holdfasts on the roots are so strong that removing them requires sandblasting; with poets, as Virgil used ivy to crown them. That one plant can simultaneously represent death, undying love, and poetry—it is a puzzle.

Fuck-You Love Poems

I study the walls of the third floor of the Capitol. They are brown with oak leaves, and the ceiling is a pale olive with yellow corn. My first thought is a savanna—the land we daily walk and work. A place for reflection. This week is the election for our next governor. Our administration will soon come to a close. I will start a new chapter of my life. I haven't told my colleagues about my plans to make art and write. I find it hard to explain myself.

Corn represents hospitality and abundance—summer barbecues, Cajun boils, and the first Thanksgiving. Acres and acres of corn form neat rows across the Midwest. But that is not the corn we eat, at least not directly. That corn fuels cars, feeds cows, provides material for bioplastics, and sweetens processed foods. That corn is a symbol of human power, wealth, and engineering. To a water treatment plant operator, corn represents fertilizer and pesticides the plant is not equipped to remove.

I study the buttery ears. I am looking for the one that stands out. The building manager says no one knows who did it. It wasn't part of any of the approved renovation plans.

First, I locate the mouse hidden at the foot of a regal male figure on the south wall. The mouse was uncovered during the recent renovations—a visual poem about feeling small. It is from a painter gone rogue at the turn of the century. The building manager had pointed it out to me, saying there were rumored menageries of elephants and horses among the acanthus leaves of the building's exterior columns. The Capitol is sometimes called "the People's House." We each want to make this space our own, to leave our mark.

Moving eastward, I find what I am looking for. Among the ears of yellow corn is a single ear dotted with reds and browns. This is the corn of the Three Sisters. Corn grown with squash and beans—grown long before the Capitol stood on this land. I think of it as a fuck-you love poem.

Invisible Unmentionable Costs

A former colleague promises me that everyone understands we are just doing our job, that animosity is only momentary. But having pushed back against the captains and their crews, having felt I could never do enough, it doesn't seem like a mist the sun can burn off. Pushing for change requires abrasion. I am sandpaper. Not *Minnesota nice.*

My anxiety seems bearable only under the dark-blue sky in the dome, with golden orchids for stars. I've already left science research, and I plan to leave politics at the end of the administration, in just a few months. As a scientist I went without sleep to run samples on mass spectrometers, traveled through a blizzard on a glacier, and took a grant application to a post office when I was in labor. I didn't yet see the cumulative effect—didn't yet believe I had limits. In politics, I hand-pumped milk into a sink during security lockdowns in the U.S. Senate. I stayed up all night to negotiate with legislators at the State Capitol. And always, I stayed calm when elected officials, advocates, and our own cabinet members yelled at me. All this has been possible, but at what cost?

Harnessing My Own Vocabulary

I open my sketch pad and flip through: the curve of a leaf, the wing of a flying lion, the faint marks left by erased lines, a lopsided acanthus leaf I gave up on, Greek keys designed to look like meandering rivers. It is a love letter full of talismans to my long-distance self. A reminder to look around. Pay attention. I'm hungry for landscape, for plants that exchange breath with me. While I've grown in this position, to continue would be self-destructive.

Cass Gilbert kept sketchbooks while he traveled across Europe. He broke buildings down into fragments: a lion's head, a scaly dragon, a doorway grill. These became part of his design vocabulary. A vocabulary flexible enough for him to create classical structures such as the U.S. Supreme Court Building, the Woolworth Building, and the George Washington Bridge.

What if I am not a drifter, but a collector? What if my anchor is not in finding where I belong but in harnessing my unique vocabulary? A vocabulary that understands how what happened millions, thousands, and hundreds of years ago created the karst landscape, the sand quarries, the rivers, and the lakes. How these landscapes are folded into our family memories and stories. How what we do today will shape the land and water far into the future. How smoke from the chaos fits into larger systems. How any change in these systems has trade-offs, and that to see these requires creating platforms for all voices to be heard. It is a vocabulary that has space for both the truth in science and the truth in our stories.

In Search of Closure

In the last week, I see the perfect reflection of the State Capitol in the Wells Fargo building downtown. The Capitol's gleaming white rectangle and dome, across the highway, reflects ourselves back at us.

When I moved back to Minnesota, I went to the fortune teller. She told me nothing about my future. Instead she held up a mirror. She mapped my personality. She assured me I could interpret my nighttime dreams. She advised me to sit down and write at the end of the day (I hadn't told her I was a writer). Based on where my husband fell in the movement of stars and moons, she told me he was happy to support my ambitions. *Seeing* myself, being seen, gave me a sense of validity beyond what any promise of tomorrow could hold.

I walk through the Capitol and look for evidence to determine: *Have I made a difference? Did my work here matter?*

I don't know until it arrives that on the final day my fingers will be chapped from feeding paper into a shredder. How without our BlackBerrys or computers there is no one to bear witness to the public's steady stream of questions and concerns. That thinking about the public's questions going unanswered will make my breath catch and my heart race.

We don't meet the new staff, though they are from the same political party. We don't pass anything on. It will be a clean start.

At a staff meeting a few months before the administration ended we filled out surveys, and now we are given a copy of everyone's answers. A memento. What do you wish someone had told you when you started working in the governor's office? *Something is always on fire, time is always against you, no one will understand your work.*

Walking out of the Capitol for the last time, I look up at the dome, my throat constricts and my stomach flips. I'm squeezed in a vice. The tug is physical. It's the umbilical cord that has tied my life to this office and to this governor twenty-four hours a day and seven days a week.

I walk out the east hallway under green grape vines. The cord pulls and pulls. I need to sit down—but if I've learned anything, it's stamina. I put one foot in front of the other and make my way across the marble floor. I push the heavy oak door open and step outside—SNAP—the cord breaks.

At the start of a drawing, the first lines look like nothing at all, but if I forge ahead, the moment of transformation will come when the sphinx is ready to fly, the flower to open, and the star to burn.

The only answers I have are in the form of questions. My transformation is just beginning. For now, I focus on how I will stand at the bus stop later. How my boys will tumble out in their snow pants and hats, how we will draw in the snow with our feet, how we will go home and play a game that comes to us in the moment. A game that needs no planning, only that we show up.

Epilogue

◆

The Blank Canvas

The Dayton administration ended in January 2019. I have a confession to make about what happens when I walk away from my career to focus on art. In truth, I have a series of confessions.

My first confession is that the bravado of my fantasy to write does not live up to my day-to-day experience. No longer held back by the limits of time, my dreams now seem barely scraps. Just wisps of ideas.

After my kids step on the bus in the morning, I keep walking. I have ants crawling on my skin, a feeling that something is wrong. I intend to walk until I pinpoint the source but turn back first.

My long-held confidence fizzles. It now appears as no more than belligerence. In science, I pushed against the edges of what we know. In politics, I pushed against the status quo. Sometimes the only woman in the room, I navigated blatant and subtle confrontations that made the high-adrenaline work all the more high-pitched. Facing resistance, I knew who I was.

Part of the fear of following my dream to write is that if I don't like it, I won't have anything else. I will have used up my fantasy.

In the months leading up to the end, headhunters called me. I thanked them for their interest and directed them to other potential candidates. My father tells me the shelf life of my professional experience is limited. It is true: I have never been more employable. I try to convince myself it is not weird to sit alone at a kitchen table and write things no one asked for, things that, even if people read and enjoy, likely won't be considered for more than a moment. I tell myself it is not weird to have no structure to my day. I tell myself it is not weird to not know how to form community outside an office.

What I did before—fighting for climate action and clean drinking water—was public service. It mattered. Carbon dioxide concentrations in the atmosphere are over 400 parts per million. The health of people and wildlife depends on action. And I am sitting alone with a pen and paper. Creative work is slippery: not made-to-order. Writing is a collaboration between the conscious and the subconscious. The questions wouldn't be worth asking if I knew the answer when I started a story or essay or poem. Creativity is about discovery, and my discoveries might not be all that interesting.

I picture former colleagues in the container of an office with a mission statement and coworkers. It sounds cozy. I thought following my heart would be romantic, but I am uncertain. Lost. Perhaps the truer confession is that without a professional title, I don't feel valid.

My phone rings, "Can you give me advice?" A leader in the new administration asks me what I think about starting a climate subcabinet. Structuring an administration is something I know about. Just like that, I have form—all sharp angles. Writing, in comparison, is like walking through a flooded marsh with mud suctioning off my boots.

I get coffee with a friend. "So, what are you doing?" she asks. The last person I explained myself to said, "I get it. You're taking a sabbatical." But that implied a break from which I would return. Finding what I seek requires more of a free fall.

"I'll spend mornings and late afternoons with the kids. I am writing. I joined a gym. I want to draw. I want to spend at least a couple of hours outdoors every day, no matter the weather. I am going to meditate daily," I pause. "And my cover story is a teaching gig at the university."

"I'm exhausted already," she shakes her head. I cringe. Her response does not reflect my intention. The point is to not operate threadbare.

In my heart of hearts, I have just two goals: First, to write. Second, to not be exhausted. I am living *Fuck carpe diem*—no racing, no multitasking.

"What do your former coworkers think?" she asks.

I had been in my office at the State Capitol when I found out I received an art grant. Turning to tell my officemate, I froze. I wanted to jump up and down. But I'd never mentioned art or writing in the office. Salty tears ran over my smile. I turned back to my keyboard. How do you tell someone who you are when they already know you as someone else?

Following the rule of no surprises, I told the governor I would not be starting a new job because I was writing a book. In our final meeting as an office, the governor said, "I've only heard one plan that sounds any good. Anna is going to be a stay-at-home mom."

The only way I know to integrate my mind and body, the past and present, science and art, is to write. Facing the blank page, I hold one hand up to keep fear at bay. The other hand plays with rhythm, scene, and lyric. I fight my way upriver to the magic state of flow—that is about the time I have to run to the school bus. It is me against myself. I have the same fight each day. I never start further along in the process.

My second confession is that off the adrenaline-fueled roller coaster, I don't sleep. At least, not well. For weeks I dream of work. The administration might be over, but all night I write briefs and take meetings. I thought stepping away would invite rest and relaxation. It does not. I cannot quench my fatigue. I lie down midday, can't exercise regularly, cancel evening plans. An acupuncturist tells me I am a flame with no wax. A burned-down candle.

Joan Didion writes, "We tell ourselves stories in order to live." The blizzard in my snow globe has settled, but I am not prepared for the open vista. The electricity in my bone marrow is not grounded. I'm trying to tell a different story, but my body is stuck in the old one.

In his book *On Intelligence,* the neuroscientist Jeff Hawkins writes, "Your brain has made a model of the world and is constantly checking that model against reality. You know where you are and what you are doing by the validity of this model." In short, my brain is a prediction machine. When I face a blank page in a metaphorical room of my own (my snowbird in-laws' kitchen table, our dining room table, a coffee shop, the public library), the pump of my heart and the firing of my neurons, based on past experience, prepare for shit to hit the fan.

My third confession is that while writing is solitary, community feeds my art. Four months after leaving the Capitol, I volunteer at the Wordplay book festival in Minneapolis. I talk to other volunteers and walk authors to their events. I'm standing at the check-in table when a tall Black man with a big smile comes in. I overhear the person working the table tell him his interviewer is sick.

He looks around, "Can someone else interview me?"

"Yes!" I shout, waving my hand in the air. He is radiant.

Jericho Brown and I sit on the balcony for the interview. We talk about his third book of poetry that just came out, *The Tradition.* We can't know yet that the book will win the Pulitzer Prize for Poetry. He tells me about combining the sonnet, the ghazel, and the blues to develop *the duplex.* "I invented a new form, I cut up all the lines I had left over from poems failed going back as far as 2004. I was splicing things together, making fragments work in ways I had never made them work before," Brown said. As in science research, the process of creation and discovery required him to forge his own path. I understand then, it isn't supposed to feel familiar when we pick up the pen to create.

At the fall equinox, my art show, *The Nature Library,* opens in downtown St. Paul. The show includes objects from nature contributed by artists, scientists, and policy experts. For each object I sew a blank book. I make books for a jar of Mississippi River water, a cross section of a tree from northern Minnesota, a core of lake mud from the suburbs, a shell from Kenya, a jar of pecans from Louisiana, an inscribed rock from Tibet, a pine cone from California, a tree core from Algeria, a wind-formed rock from Antarctica, and a rock picked up by a research submarine near the Bermuda Triangle.

The first page of each book holds the contributor's story. The rest of the pages offer space to visitors. In looped cursive, uniform small print, and sprawling script, visitors write about loved ones long gone, childhood, gifts given and received, cancer treatments, immigration, and trapeze arts.

The opening reception includes writers, artists, my students, science colleagues, the former governor, political colleagues, and others. A friend from college plays fiddle. My son restocks the plates of cookies he baked. I joke with my husband that no one knows it is my secret early fortieth birthday party.

"You know what you are doing?" a local author and creative writing professor says to me.

"What am I doing?"

"Teaching," she smiles. "You should teach writing." I am uncertain I know enough to teach, but I smile. I feel seen.

I look down at the postcards visitors can take as souvenirs. I printed the sage leaves, fern fronds, and pine needles in ink with my six-year-old. A question runs along the border: "What is the story you read in the landscape?" The artist Austin Kleon advises a cycle of learning and teaching. When you learn something, turn around and share it. The next day I submit a proposal to teach a class at the Loft Literary Center.

My fourth confession is that as a pandemic spreads across the globe, I get a diagnosis. Something is actually wrong. Though the problem is not visible to the eye, my body is broken. Our city is broken, too. Only five miles from our house, police kill George Floyd on Memorial Day. By the end of that week every gas station is an open-mouth cry of burned-down rubble. The virus is everywhere. We tell ourselves it will be a week, then we say a month, and then time loses texture.

My pause is now part of a global pause. My weird a global weird. The shutdown normalizes what I am doing—everyone must rethink life, rebalance. But telling myself I could go back to the office was what made my desertion safe. I left the planet to free-float, and while I was gone the planet turned upside down, shook out its cities and pockets and refused to put anything back in place. Everyone is free-floating.

The night of the 2020 summer solstice I wake with chills like a hundred knife blades in my collar bone. The windows are open, the house warm. My teeth chatter. I try a hot shower, a sweatshirt, a wool hat, wool slippers, a hot-water bottle, cheese melted on toast. It is summer, and I can't stop shivering. I check my Fitbit—my heart rate has been in the forties for hours, well below my usual resting rate.

I call the doctor, who tells me I don't have Covid. This sounds ridiculous. She sends me for a blood test, and, based on the results, prescribes a daily hormone replacement. I will take this for the rest of my life. The typical symptoms of an underactive thyroid are depression, fatigue, and weight gain. "Could you have it mixed up?" I ask. I am enthusiastic, active, always up to something.

Our city, the environmental movement, the university, the arts community, everyone on my social media feed are starting a deeper and more urgent reckoning on race. My diagnosis feels inconsequential. To be sick with anything but the virus is a distraction. The medicine changes my senses of smell and taste. Waves of energy push me out on long runs. It's hard to eat. Sleep is different, my body is different, the world is different. I'd turned forty in March, the pandemic an advancing shadow, but it is only now that I feel over the hill. *Forty is not the new twenty.* Life is actually short.

My seven-year-old draws a picture of me with a butterfly over my throat that lifts to reveal a smaller butterfly and then smaller and smaller butterflies: nesting dolls of thyroids. The question nags at me whether my *Fuck carpe diem* mantra is just a response to my heart and metabolism not working properly. I wonder, if I hadn't been so depleted, could I have laughed off the stress of politics?

Staring at my thyroid data I understand Covid conspiracy theorists, climate deniers, and everyone who fought the environmental regulations I worked so hard for. Despite clear numbers from my blood tests, I don't want to believe the diagnosis. To do so requires rewriting my story. I'm a chemical reaction of anger and disappointment, slow to come around.

As months go by, the sensory deprivation of the pandemic deepens. I blast Lady Gaga and dream of crowded stadium concerts, neither of which I liked before. I don't know what I am looking for, but I seek wearing bright-red lipstick, getting a belly button piercing, making a dress from bike tires, hiking at night without lights. The poet Jane Hirshfield says, "Poetry's task is to reunite us with the senses, with our hungers, with our emotions, with our physical connection to the bones inside our hands." I am hungry, and, in fear and isolation, I lose poetry over and over.

My fifth confession is that my cover story turns out to be real. On my forty-first birthday we meet friends for a bonfire on a frozen lake. This is the grace of Minnesota winter: ephemeral public spaces. Our kids have just returned to school in person, with their faces hidden behind masks. Our friends have a nephew on a ventilator and don't know if he will live. But we can gather outside in layers and layers of fleece and down and wool. The flames from the bonfire make our faces glow even if our feet slosh in the melting snowpack.

The friend who works as a chiropractor asks what I have been doing. I tell him about writing and say, "but I also have a cover story." As if that is an apology or a permission slip.

He says, "Totally. Being a chiropractor is my cover story." I ask what his cover story is for. He looks at me like, if I told you, I would have to kill you. It takes only a moment before I laugh. The sequestering of what we love most is a survival instinct.

I know he loves his patients. He uses his whole body to talk about work. Being a chiropractor is not a front or a lie. It's just that he contains so much more. Our jobs too often fill in as a shorthand for our essence. We feel we know someone, or know enough about them, or know how to categorize them, when we know how they make money.

Over the past two years, I've built a constellation of paid work. It is a lot easier to tell people about this than about writing. Often it is all I tell. When someone asks me, "And what do you do?" everyone seems most at ease if I respond with something that fits into an economic bucket. I say, "I teach" or "I consult," and they hear, *I am a responsible adult.*

For my teaching, I make syllabuses for public policy graduate students and prepare mock legislative hearing questions for a class with law students and research scientists. I fit what had been chaotic in politics into tidy weekly lectures and term papers. With distance and the unrelenting questions and ideals of my students, I gain perspective.

My students form a diorama that plays out the paradox of advocacy and progress. In staged debates, those on the side arguing for environmental regulation read facts in monotone voices, throwing out proof in acres of pollution, percentages of lakes, and parts per million of chemicals. The numbers rattle around the room with no place to land. The side arguing against regulation tells stories ripe with emotion. Speaking as turkey farmers or county commissioners, students talk about how regulations would harm their businesses and, by connection, their families.

I try to convince my students that the point is not to prove you are right—it is to build trust with others, particularly those who don't agree with you. Compromise. Work across parties. See the other side not as a force of evil and destruction but as people. Find points of overlap and build coalitions. Take the steps forward you can take with others. The students respond with a call for immediate and comprehensive action.

It is another year and a half before the chiropractor and I meet to write together. He finally told me: he is a poet. It's cool for September, but the day warms as our pens move across the page. We sit at a table in his yard, two responsible adults, playing with words, following our hearts, no path laid to reach any goal.

My final confession is that when my attempts to make a difference in policy and politics failed or I moved on, the campaigns did not end. The baton was handed off. Others took over. The clean energy bill I worked on for Senator Franken passed in bits and pieces over the next decade. The new administration in Minnesota dedicated more staff and resources to environmental justice. The drinking water protection rule caught in a game of chicken between Governor Dayton and the legislative committee chairs went into effect. The state of Minnesota transferred sacred land that had been operated as a state park back to the Upper Sioux Community. The momentum continues. Farmers form groups to lead on climate. Environmental organizations, government, philanthropy, and new media outlets evolve to include more voices and collaboration with more communities.

I keep my cover story for both practical and meaningful reasons. I maintain my art practice. I had thought that making the choice to write, *choosing art*, was righteous. I realize now that what was actually spiritual was the act of choosing. No choice was more worthy than another. Seeing the crisp edges of mortality and letting go of parts of myself is brave. By not having it all, I actively become myself.

Action is collective through both time and space. All of us are important and responsible. Some of us are loud. Some quiet. Our actions result in immediate impacts or plant seeds that germinate over the long term. Sometimes our work sets someone up to succeed later. Some of us work in the system and some work outside it. Some break things apart. Some barricade. Some build. Some of us are afraid. Some angry. Some overwhelmed. Some hopeful. Some chase winter. Some start businesses. Some talk to neighbors. Some talk to strangers. Some read. Some write. Some listen. Some teach. Some plant trees. It takes all of this, all of us.

No matter how much paint we throw at the canvas, the future is always blank. Yet to be lived.

The future is all of ours to make.

Acknowledgments

The act of putting words on a page is solitary, but what comes out is brewed, ruminated, and fed by community. Fragments of essays first came out in my yearlong *Science Love Letter* postcard project with Natalie Vestin. A couple of essays started in writing classes with Catherine Imbriglio and Elizabeth Taylor decades ago. Regular phone calls with Bronwen Tate and Ranae Hanson to discuss our writing projects have meant the world. Texting sentences or sending sections of the book to my sister Hope Henderson has kept me going. Thank you to my agent, Malaga Baldi, for believing in me and supporting me. The book grew and developed with the support of Kristian Tvedten and all the wonderful team at the University of Minnesota Press.

Portions of some of these essays have been published, sometimes in different forms, in *The Rumpus* ("How Strangers Form Constellations That Make Sense of the World"), *Kenyon Review* ("Dear Spartina"), *River Teeth* ("View from the Lactation Room at the White House"), *The Common* ("Saying Yes to the Mountain"), *Seneca Review* ("From Scientist to Animal"), *The Doctor T. J. Eckleburg Review* ("The Secret Mice of the Smithsonian"), *Water~Stone Review* ("Whatever Discomfort, Find Beauty"), *Cleaver Magazine* ("Packing for an Overnight at the State Capitol"), *No Contact* ("Thank You for Lobbying Our Office"), *Identity Theory* ("Fuck Carpe Diem"), and *Raritan Quarterly* ("The Irrational Logic of Our Bodies on the Landscape").

Thank you to my amazing writing group I met through Grub Street: Emily Green, Daniel Cohen, Sylvia Baedorf Kassis, and Rose Darline Darbouze. Thank you to the amazing group I met through Peter Geye's workshop at the Loft Literary Center: Cheryl Bailey, Meagh Decker, Megan Smith, Tom Sebanc, and Tariq Samad. Thanks to my thoughtful workshop group at Breadloaf Environmental Writers Workshop led by Kazim Ali. Thanks to Carolyn Williams-Noren for collaborating on public art at her Little Poetry Library. Thank you to Patricia Weaver Francisco and Katrin Schumann for help on the manuscript and figuring out book publishing.

Thanks to the Loft Literary Center and all the students who have taken classes with me. I learn so much from you. Thanks to my colleagues and public policy students at the University of Minnesota Humphrey School of Public Affairs and the law students and scientists from the Mitchell Hamline School of Law who have participated in the Expert Witness Training Academy. Teaching policy was essential for my learning.

I am grateful to the many individuals and organizations who helped me become a scientist, enter policy work, and write this book. While their support was essential, the content in this book does not reflect their views, and any inaccuracies or opinions are mine and mine alone. Thank you to the Limnological Research Center and all its amazing team. Thank you to the American Association for the Advancement of Science (AAAS) and the American Geosciences Institute for their support. Thank you to all my wonderful colleagues in Senator Al Franken's office, and especially for the mentorship of Ali Nouri. Thank you to my colleagues in Minnesota state government and Governor Dayton's office. These public servants are the bravest entrepreneurs and change makers.

This book could not have happened without the support and insights of all those who read passages, gave feedback, and talked with me about the book, including Amber Larson, Almea Matanock and Nat Chakeres, Kristina Brady Shannon, Amy Myrbo, Kate Moos, Brooke White, Tenzin Dolkar, Galen Treuer, Susan Moore, Jeremy Bellay, Krisanne Dattir, Nicole Rom, Karine Rupp Stanko, Jessica Deutsch, Soumaya Belmecheri, Ramona Treuer,

Heather Graham, Laurence Bird, Amy Fredregill, Barry Taylor, and Tovah Bodner. Much thanks for emotional support from Emily Esch, Victor Sta. Ana, Michelle Higa Fox, Hill Chau, and Susan Wong. Much thanks for support to work on the book from the Minnesota State Arts Board, a residency at the Madeline Island School of Arts, the Loft Mentor Series Program, and a residency at Everwood Farmstead. Each opportunity broadened not only my creative work but also my creative community.

All my heart and appreciation to Dan, Silas, Walter, Sky, and Waffle.

ANNA FARRO HENDERSON is a Canadian American scientist and writer. She was an environmental policy adviser to Minnesota Senator Al Franken and Governor Mark Dayton and is now a fellow at the Institute on the Environment at the University of Minnesota, working in climate advocacy, and an instructor at the Loft Literary Center. She lives with her family in Minnesota, where she makes daily visits to the Mississippi River.